S0-CPS-412

ELASTICITY AND ANELASTICITY OF METALS

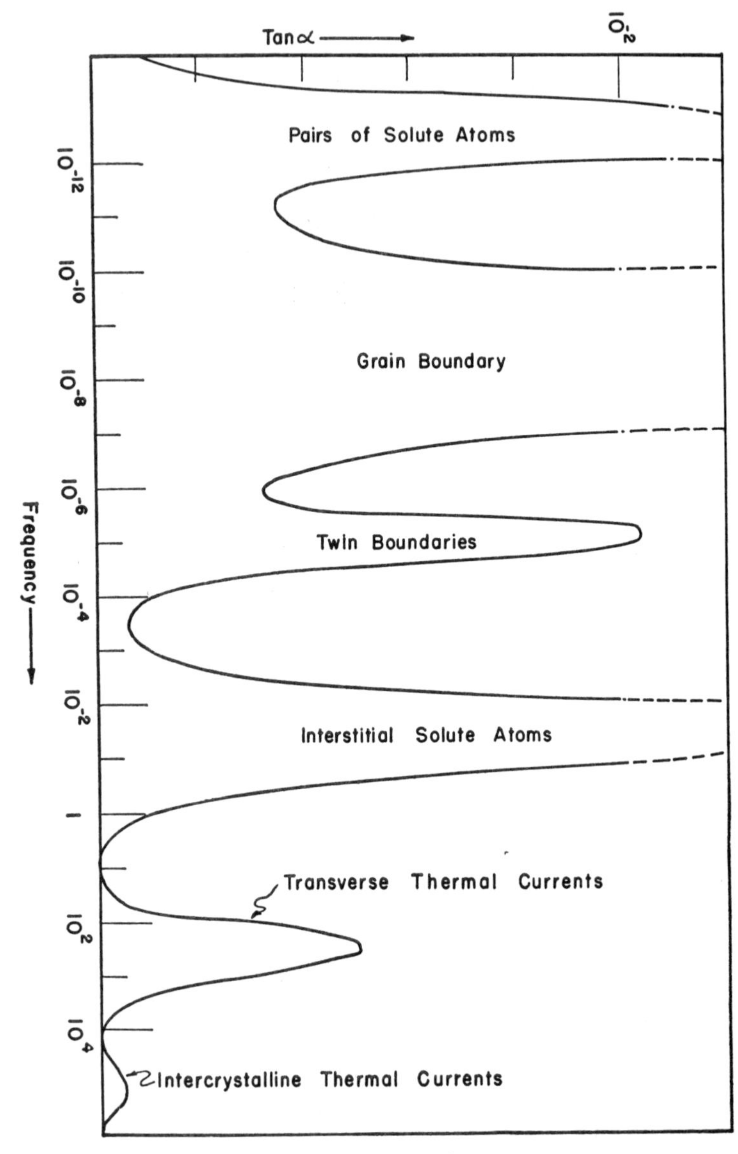

A TYPICAL RELAXATION SPECTRUM

ELASTICITY
AND
ANELASTICITY
OF
METALS

By
CLARENCE ZENER

THE UNIVERSITY OF CHICAGO PRESS
CHICAGO & LONDON

THE UNIVERSITY OF CHICAGO PRESS, CHICAGO & LONDON
The University of Toronto Press, Toronto 5, Canada

Published 1948. Fifth Impression 1965. Printed in the United States of America

FOREWORD

METALLURGY is an ancient profession and one in which science has played an increasingly important part. Metallurgists had acquired a vast body of empirical knowledge and practical skill long before the development of the formal sciences requisite to its intellectual understanding. For centuries the chemist learned from the metallurgist; but in the nineteenth century the tables were turned: the acceleration of metallurgical progress came from the application of chemical science. The methods of analysis and the chemical preparation of pure materials made possible, for the first time, realistic studies of the effect of composition and impurities on technical behavior and properties. The twentieth century has seen the growth of X-ray diffraction and other methods of the physicist and their application to metals. It seems likely that the important developments of the next generation will come about by the application of the principles of physics, chemistry, and mathematics to the study of structure, properties, and behavior of metals.

Late in 1945 the University of Chicago established three institutes for research in fields involving more than one established discipline. One of these was the Institute for the Study of Metals. Just as in the new fields of nucleonics and biophysics, so also in the old field of metallurgy there seemed to be need for an organization in which, under a common roof and exposed to mutual stimulation, physicists, chemists, and more specialized scientists could attack a broad problem. It is mutually advantageous for the physicists to learn and utilize some of the large body of rather empirical metallurgical information and for the metallurgist to employ the rigor of the methods of the physicist and physical chemist.

The advance of knowledge in any field is closely related to its literature, for personal contact is too limited in both time and space. Soon after the Institute was established, it became obvious that its aims could be furthered by monographs on various aspects of the science of metals, and a series was established under the joint auspices of the Institute and the University of Chicago Press. Works by any author are eligible, provided that they deal in a competent manner with some aspect of metal science or closely related fields. Textbooks and discussions of industrial "know-how" are excluded because excellent outlets for such material already exist. Three kinds of monographs are to be encouraged: critical reviews of particular topics that seem ripe for research—the kind of work that will generally be written by a scientist as he makes a detailed survey of the status of a field before undertaking intensive original work on it; extended summaries written after prolonged work in a specialized field at the time when a scientist wishes to record his work in perspective as a basis for future investigations by himself and others; and, lastly, books discussing the application of particular scientific principles or techniques to metallurgy and related subjects.

The present book, *Elasticity and Anelasticity of Metals*, by Clarence Zener, deals with a subject on the border line of metallurgy, physics, and engineering. Very largely as a result of Dr. Zener's work, internal friction has become an important tool in the study of the structure and behavior of metals. There are already many different types of anelastic deformation known; and it seems likely that the acoustic spectrum will come to occupy a position in solid-state studies comparable to that possessed by the optical spectrum in the theory of atomic and molecular structure. Though some of the data in this book have appeared before, this is the first time that a connected and thorough treatment of the whole subject has been written, and it should prove of considerable value to physicists, metallurgists, and engineers alike. The section on microelasticity, hitherto unpublished, will be found to have considerable relevance to the stability and transformation of metallic phases. This is a work that the committee is particularly happy to sponsor as the first member of the series.

CYRIL STANLEY SMITH, *Chairman*
CHARLES S. BARRETT
EARL A. LONG

Committee on Monographs
Institute for the Study of Metals

PREFACE

THAT metals manifest deviations from perfect elastic behavior, even at small stress levels, has been known for over a century. Only during the past decade, however, has marked progress been made in the interpretation of these deviations in terms of the microstructure of the metal. Although not many years ago these deviations were investigated only by erudite students of metals and were regarded by metallurgists as interesting but quite useless, today they are being studied to answer very definite questions regarding the kinetics of changes in the microstructure. In preparing this monograph the author has attempted to present in a logical manner the science of the nonelastic behavior of metals at low stress levels, to review all the pertinent work in this field, and to point out the role which studies of this behavior will play in the future development of metallurgical science. Anticipating that all who may want to profit from this monograph will not wish to follow the formal development of the theory, the author has endeavored to write in such a manner that the formal developments may be omitted by an intelligent reader without loss of a physical understanding of the subject. The elastic and nonelastic properties of metals are so interrelated that it has seemed inappropriate to discuss the latter at length without at least a brief introduction to the former.

Nonelastic behavior carries with it a connotation of a permanent set after removal of all stress. Although metals behave nonelastically even at low stress levels in the sense that stress and strain are not single-valued functions of each other, e.g., strain may lag behind stress in periodic vibrations, no permanent set remains after removal of all stresses. It has therefore seemed appropriate to introduce a new term that is free of this connotation of permanent set. *Anelasticity* has been chosen to denote that property of solids in virtue of which stress and strain are not single-valued functions of one another in that low stress range in which no permanent set occurs and in which the relation of stress to strain is still linear.

The author is indebted to the Office of Naval Research for its continued support of research upon the anelasticity of metals, a support which has aided materially in the preparation of this monograph.

TABLE OF CONTENTS

INDEX

INTRODUCTION

THE modern theory of elasticity may be traced to a law formulated by Hooke[1] in the seventeenth century: "Ut tensio sic vis." According to this law, if a "force" is applied to a body, the resulting deformation is proportional thereto.

The "force" in Hooke's law may be an actual force or any related quantity, such as torque, couple, etc. The deformation is usually so defined that the work done upon the body when the deformation changes by a small amount is equal to the force *times* the increment of the deformation. Thus, if F is the force, the deformation, D, is so defined that

$$\delta \text{ work} = F \cdot \delta D .$$

Thus, for a force, torque, or couple, D refers to displacement, angle of twist, and angle of bend, respectively.

The mathematical formulation of Hooke's law is

$$MD = F , \tag{i}$$

where the coefficient M, known as the "elastic modulus," is independent of the force F. The modulus does, however, depend upon the type of the applied force. Thus M may refer to the tensile modulus, the torsion modulus, or the bending modulus.

As originally formulated, Hooke's law is applicable only to quasi-static forces, i.e., to forces very slowly applied. This limitation is evident when it is realized that the deformation, D, in Hooke's law is influenced by distortions throughout the entire body. The deformation cannot, therefore, attain its value associated with the applied force until, in effect, every part of the body has been notified of the force and the resulting distortions have been signaled back to the region of application of the force. It is evident that the equilibrium deformation cannot be established in a time less than the time required for an elastic wave to travel from the region of application of the force to the most remote part of the body and back again. As an example, an impulsive transverse force applied to a taut string produces a rectangular wave which extends rapidly toward the supports. In place of Hooke's law, the relation between force and displacement in this case is

$$\frac{dD}{dt} = \text{Constant} \times F .$$

1. Robert Hooke, *De potentia restitutiva* (London, 1678).

This equation is valid until waves have been reflected back from the end supports. Hooke's law may not be applicable even after elastic waves have been reflected from the most remote parts of the specimen. The elastic waves may continue to travel to and fro. In other words, the system may be set into vibration by the force. A force may be regarded as applied in a quasi-static manner, with no attendant vibrations, only if the time taken to apply it is long compared with the period of the slowest mode of vibration. In order that the original formulation of Hooke's law may be applicable, the force must therefore be applied the more slowly, the larger the body.

During the early part of the nineteenth century Hooke's law was generalized in such a manner as to remove the restriction that forces must be applied in a quasi-static manner. In this generalization the body is considered as partitioned into a great number of elementary regions of very small extent. The distortion of each region is then determined solely by the forces acting upon it by the neighboring regions, irrespective of how rapidly these forces change. In effect, one regards the elementary regions as so small that all forces change only a negligible amount during a time interval equal to the period of their slowest mode of vibration. The change in shape of the elementary regions may be completely described in terms of certain quantities called "strain components." The generalized Hooke's law then relates the strain components to the force components acting on the elementary regions or, more properly, to the surface density of the force components, known as "stress components." The variation of the stress throughout the body is determined by this generalized Hooke's law together with the equations of motion of each elementary region.

As an example of the above generalization, reference will be made to the simple case of a force applied to one end of a wire, the other end of which is rigidly fixed. When a force F is applied to the end of the wire, the change in length of the wire will be proportional to F only if the force is applied quasi-statically. According to the above-outlined method, one focuses attention on an element of the wire of original length δX. Since δX is arbitrarily small, the acceleration of the elementary region will be finite only if the forces acting upon its two ends are equal in magnitude and opposite in direction. These forces will be denoted by f. The strain e of the element is defined in terms of the original and final coordinates, X and U,

$$e = \frac{dU}{dX} - 1 . \tag{ii}$$

Then

$$f = \text{Constant} \times e \tag{iii}$$

is the generalized Hooke's law appropriate for this special case. The variation of the force f along the specimen is given by

$$\frac{m\,d^2U}{dt^2}=\frac{df}{dX}, \tag{iv}$$

where m is the mass per unit original length. Equations (ii)–(iv), together with the appropriate boundary conditions, completely specify the response of the wire to a force F applied at its end in an arbitrary manner.

The classical theory of elasticity is essentially the development of all the consequences arising from the application of Hooke's law to elementary regions. Many excellent treatises[2] have been written on this classical theory, as well as an extensive history[3] of its development. In the discussion of elasticity presented in this monograph we are concerned only with the coefficients which relate stress to strain and with the dependence of these coefficients on the various physical parameters.

The authors of the classical theory of elasticity were under no illusion as to the lack of strict applicability of their fundamental assumptions to real solids. The value of the classical theory lies not in its precise description of the behavior of solids under applied forces but in its description of this behavior with sufficient accuracy for most practical purposes.

According to the classical theory of elasticity, stress and strain are uniquely related. Also applied force and deformation are uniquely related, provided that the force is applied in a quasi-static manner, i.e., provided that it is varied so slowly that no vibrations result. Weber,[4] as early as 1825, found, in his investigations upon galvanometer suspensions, small deviations from perfect elasticity. Upon release of the couple, the suspension did not at once return to its zero-point but approached this zero-point only asymptotically. Such behavior he called an "elastic aftereffect" (*elastische Nachwirkung*), an early summary of which was presented by Auerbach.[5] Other effects which are contrary to the theory of elasticity but which are shown by real solids are internal friction, stress relaxation, and variation of modulus with frequency of measurements. All

2. E.g., A. E. H. Love, *Mathematical Theory of Elasticity* (4th ed.; Cambridge: Cambridge University Press, 1927); R. V. Southwell, *Theory of Elasticity* (London: Oxford University Press, 1936).

3. I. Todhunter and K. Pearson, *History of the Theory of Elasticity* (Cambridge: Cambridge University Press, 1886).

4. W. Weber, "Über die Elastizität der Seidenfaden," *Poggendorff's Ann.*, XXXV (1834), 247; and "Über die Elastizität fester Körper," *ibid.*, XXIV (2d ser., 1841), 1.

5. F. Auerbach, "Elastische Nachwirkung," *Winkelmann's Handbuch der Physik*, I (Breslau, 1891), 769–831.

such effects are different manifestations of the lack of uniqueness of the relation between stress and strain. The property of a solid in virtue of which stress and strain are not uniquely related in the pre-plastic range is called "anelasticity." In the discussion of anelasticity presented in this monograph we are concerned primarily with the physical origin of the various anelastic effects.

PART ONE

ELASTICITY OF METALS

I

FORMAL RELATIONS BETWEEN STRESS AND STRAIN

NO ESSENTIAL additions have been made to the formal theory of the elasticity of crystals since the publication of Voigt's[1] excellent treatise. This formal theory will therefore be only briefly reviewed, with no attempt at completeness.

A. STRAIN

The concept of strain components has been introduced in order that small deformations of a body may be uniquely specified. In preparation for a definition of the strain components, we shall first define the displacement vector. As a result of deformation, a particle which was originally at x, y, z will now find itself at $x + U$, $y + V$, $z + W$. The vector (U, V, W) is called the "displacement vector," and its components are functions of x, y, and z. We now confine our attention to a small element of volume, as indicated in Figure 1, and regard the origin of coordinates as lying within the element before deformation.

The displacement of every part of this small element may then be expressed in terms of twelve constants, namely, as

$$U = U_0 + \frac{\partial U}{\partial x} x + \frac{\partial U}{\partial y} y + \frac{\partial U}{\partial z} z\,, \quad \text{etc.,}$$

where the derivatives are taken at the origin. The vector (U_0, V_0, W_0) gives the displacement of the origin of coordinates. The displacement of every part of the body with respect to the origin, (U', V', W'), is then given in terms of nine constants, namely, their nine partial derivatives.

Under certain important conditions a more direct physical interpretation may be ascribed to nine other constants which may be formed by a linear combination of the above nine. These are as follows:

$$\left.\begin{aligned}
e_{xx} &= \frac{\partial U}{\partial x}\,, \qquad e_{yy} = \frac{\partial V}{\partial y}\,, \qquad e_{zz} = \frac{\partial W}{\partial z}\,; \\
e_{yz} &= \frac{\partial W}{\partial y} + \frac{\partial V}{\partial z}\,, \qquad e_{zx} = \frac{\partial U}{\partial z} + \frac{\partial W}{\partial x}\,, \\
e_{xy} &= \frac{\partial V}{\partial x} + \frac{\partial U}{\partial y}\,, \qquad \omega_x = \frac{\partial W}{\partial y} - \frac{\partial V}{\partial z}\,, \\
\omega_y &= \frac{\partial U}{\partial z} - \frac{\partial W}{\partial x}\,, \qquad \omega_z = \frac{\partial V}{\partial x} - \frac{\partial U}{\partial y}\,.
\end{aligned}\right\} \tag{1}$$

1. W. Voigt, *Lehrbuch der Kristallphysik* (Berlin: Teubner, 1910).

These nine constants may be given direct physical interpretations when they are small-order quantities, i.e., very small compared with unity. Under this condition the deformation relative to the origin may be considered as taking place in two steps. The first deformation is represented by the displacement vector,

$$\left.\begin{aligned} U_S &= e_{xx}x + \tfrac{1}{2}e_{xy}y + \tfrac{1}{2}e_{zx}z\,, \\ V_S &= \tfrac{1}{2}e_{xy}x + e_{yy}y + \tfrac{1}{2}e_{yz}z\,, \\ W_S &= \tfrac{1}{2}e_{zx}x + \tfrac{1}{2}e_{yz}y + e_{zz}z\,, \end{aligned}\right\} \tag{2}$$

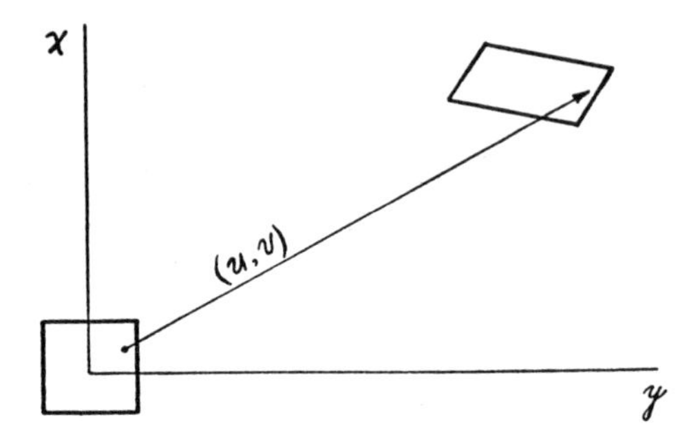

FIG. 1.—Illustration of displacement vector in two dimensions

in which the coefficients form a symmetrical matrix. The second deformation is represented by the displacement vector,

$$\left.\begin{aligned} U_A &= \qquad\qquad -\tfrac{1}{2}\omega_z y + \tfrac{1}{2}\omega_y z\,, \\ V_A &= \tfrac{1}{2}\omega_z x \qquad\qquad -\tfrac{1}{2}\omega_x z\,, \\ W_A &= -\tfrac{1}{2}\omega_y x + \tfrac{1}{2}\omega_x y \qquad\qquad , \end{aligned}\right\} \tag{3}$$

in which the coefficients form an antisymmetrical matrix.

It may readily be shown that the displacement vector (U_S, V_S, W_S) represents a distortion for which a particular set of coordinate axes attached to the body do not change direction during the deformation. These axes are known as the "principal axes" of the distortion. The six quantities, e_{xx}, e_{yy}, e_{zz}, e_{yz}, e_{zx}, and e_{xy}, are known as the "strain components." The first strain component may be regarded as the relative change in length of a line originally parallel to the x-axis; and an analogous interpretation may be given the second and third strain components. The strain component e_{yz} may be regarded as the change in the angle between the y- and the z-axis during deformation if these axes are regarded as fixed to the body during the distortion; and a similar interpretation may be given to e_{zx} and e_{xy}.

The numerical values of the strain components depend upon the orientation of the coordinate axes to which they refer. It is often desired to express the strain components referred to one set of axes in terms of the strain components referred to a second set of axes. Let the transformation between the two sets of axes be specified by the following scheme:

$$\left.\begin{array}{c|ccc} & x & y & z \\ \hline x' & l_1 & m_1 & n_1 \\ y' & l_2 & m_2 & n_2 \\ z' & l_3 & m_3 & n_3 \end{array}\right\} \quad (4)$$

Thus

$$x' = l_1 x + m_1 y + n_1 z$$

and, conversely,

$$x = l_1 x' + l_2 y' + l_3 z' .$$

The transformation equations between the two sets of strain components are then obtained by transforming the set of equations (2) into one containing U'_S, V'_S, W'_S, and x', y', and z'. Thus we multiply the three equations by l_1, l_2, and l_3, respectively, and then form their sum, thereby obtaining U'_S in terms of x, y, and z. We now transform x, y, and z into x', y', and z'. The result is

$$e_{x'x'} = e_{xx}l_1^2 + e_{yy}m_1^2 + e_{zz}n_1^2 + e_{yz}m_1n_1 + e_{zx}n_1l_1 + e_{xy}l_1m_1 \text{ , etc.} \quad (5)$$

$$\left.\begin{aligned} e_{y'z'} = 2\, e_{xx}l_2l_3 + 2\, e_{yy}m_2m_3 + 2\, e_{zz}n_2n_3 + e_{yz}(m_2n_3 + m_3n_2) \\ + e_{zx}(n_2l_3 + n_3l_2) + e_{xy}(l_2m_3 + l_3m_2) \text{ , etc.} \end{aligned}\right\} \quad (6)$$

It may readily be shown that the displacement vector (U_A, V_A, W_A) represents a deformation in which the body suffers a rotation as a rigid body about its origin. The quantities ω_x, ω_y, and ω_z are the three components of the rotation vector and therefore transform with a rotation of coordinate axes according to scheme (4).

B. STRESS

The concept of stress has been introduced in order that we may specify completely the traction which acts across all planes passing through a given point within a solid. In preparation for the definition of stress components we shall consider the forces acting across an element of surface within the body. In particular, we shall consider the forces with which the material on one side of an element of internal area, dA, acts upon the material on the other side. This force will be proportional to dA, and may therefore be written as $\mathbf{T}dA$, as illustrated in Figure 2. Since the sign of $\mathbf{T}$ will depend upon which side we are considering as exerting the force, ambiguity will be removed by assigning to $\mathbf{T}$ the subscript $\mathbf{n}$, which denotes the unit vector normal to dA pointing toward the region which we regard as exerting the force. This same notation also specifies the orientation of the element of area dA.

Upon applying to the tetrahedron of Figure 3 the condition that the total force acting upon a body in equilibrium must be zero, one finds that the force density, $\mathbf{T_n}$, may be expressed in terms of the force densities acting across the *xy*-, *yz*-, and *zx*-planes. If l, m, and n refer to the direction cosines of $\mathbf{n}$, one obtains for the components of $\mathbf{T_n}$,

$$\left.\begin{aligned} X_n &= lX_x + mX_y + nX_z \,, \\ Y_n &= lY_x + mY_y + nY_z \,, \\ Z_n &= lZ_x + mZ_y + nZ_z \,. \end{aligned}\right\} \qquad (7)$$

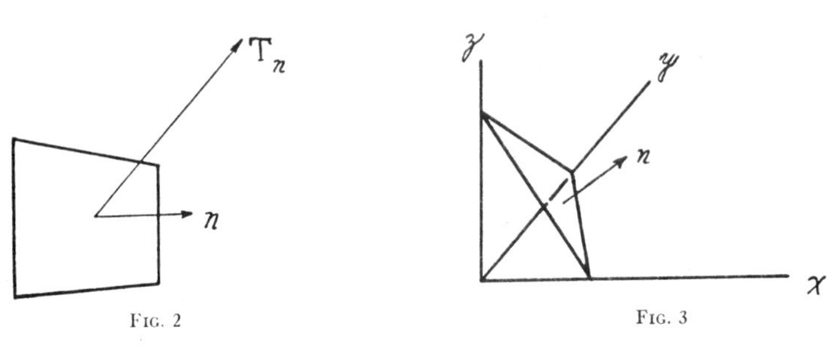

FIG. 2.—Illustration of traction
FIG. 3.—Method of resolving traction across an arbitrary plane

The components of force density X_x, . . . , are called "stress components." The number of independent stress components may be reduced to six upon applying to an elementary cube the condition that the total moment acting thereon must be zero. One finds

$$Y_z = Z_y \,, \qquad Z_x = X_z \,, \qquad X_y = Y_x \,. \qquad (8)$$

The independent stress components may therefore be taken as X_x, Y_y, Z_z, Y_z, Z_x, and X_y. The traction across an arbitrary plane may be expressed in terms of these six stress components through equations (7) and (8).

The numerical value of the stress components will depend upon the particular reference Cartesian axes. The relation between the stress components referred to different axes may be obtained by the use of equations (7). We shall consider two sets of axes related by scheme (4). The quantity $X'_{x'}$ is obtained by letting $\mathbf{n}$ be directed along the x'-axis, multiplying the first of equations (7) by l_1, the second by m_1, the third by n_1, and forming the sum. We obtain

$$X'_{x'} = l_1^2 X_x + m_1^2 Y_y + n_1^2 Z_z + 2m_1 n_1 Y_z + 2n_1 l_1 Z_x + 2l_1 m_1 X_y \,,$$

with two analogous equations for $Y'_{y'}$ and $Z'_{z'}$; and

$$\begin{aligned} X'_{y'} = l_1 l_2 X_x + m_1 m_2 Y_y + n_1 n_2 Z_z + (m_1 n_2 + m_2 n_1)\, Y_z \\ + (n_1 l_2 + n_2 l_1)\, Z_x + (l_1 m_2 + l_2 m_1)\, X_y \end{aligned} \qquad (9)$$

with two analogous equations for $Y'_{z'}$ and $Z'_{x'}$.

C. RELATION BETWEEN STRESS AND STRAIN

We have seen that the deformation in any elementary region is completely specified by the six strain components, e_{xx}, e_{yy}, e_{zz}, e_{yz}, e_{zx}, and e_{xy}. We have further seen that the traction across an arbitrary plane passing through the elementary region is specified completely by the six stress components, X_x, Y_y, Z_z, Y_z, Z_x, and X_y. The generalized form of Hooke's law states that the strain components are linear functions of the stress components and, conversely, that the stress components are linear functions of the strain components. Thus, on the one hand, we have

$$\left.\begin{aligned} e_{xx} &= s_{11}X_x + s_{12}Y_y + s_{13}Z_z + s_{14}Y_z + s_{15}Z_x + s_{16}X_y\,, \\ e_{yy} &= s_{21}X_x + s_{22}Y_y + s_{23}Z_z + s_{24}Y_z + s_{25}Z_x + s_{26}X_y\,, \\ &\cdots\cdots\cdots\cdots\cdots\cdots\,, \\ e_{xy} &= s_{61}X_x + s_{62}Y_y + s_{63}Z_z + s_{64}Y_z + s_{65}Z_x + s_{66}X_y\,; \end{aligned}\right\} \tag{10}$$

and, on the other hand,

$$\left.\begin{aligned} X_x &= c_{11}e_{xx} + c_{12}e_{yy} + c_{13}e_{zz} + c_{14}e_{yz} + c_{15}e_{zx} + c_{16}e_{xy}\,, \\ Y_y &= c_{21}e_{xx} + c_{22}e_{yy} + c_{23}e_{zz} + c_{24}e_{yz} + c_{25}e_{zx} + c_{26}e_{xy}\,, \\ &\cdots\cdots\cdots\cdots\cdots\cdots\,, \\ X_y &= c_{61}e_{xx} + c_{62}e_{yy} + c_{63}e_{zz} + c_{64}e_{yz} + c_{65}e_{zx} + c_{66}e_{xy}\,. \end{aligned}\right\} \tag{11}$$

The quantities s_{11}, s_{12}, . . . , are called the "elastic constants," the quantities c_{11}, c_{12}, . . . , the "elastic coefficients." It may readily be verified that the matrix of the elastic constants, $\|s\|$, and the matrix of the elastic coefficients, $\|c\|$, may be regarded as the reciprocal of each other in the sense that

$$\|s\| \cdot \|c\| = \|1\|\,, \tag{12}$$

where $\|1\|$ is the unit matrix; thus

$$\sum_{k=1}^{6} s_{jk}c_{kj'} = \begin{cases} 1, & j' = j\,, \\ 0, & j' \neq j\,, \end{cases} \tag{13}$$

and

$$\sum_{k=1}^{6} c_{jk}s_{kj'} = \begin{cases} 1, & j' = j\,, \\ 0, & j' \neq j\,. \end{cases} \tag{14}$$

The thirty-six elastic coefficients are not all independent. In order to derive those relations that hold for all crystals, we observe that an increment of the strain-energy density, U, is given by

$$dU = X_x de_{xx} + Y_y de_{yy} + Z_z de_{zz} + Y_z de_{yz} + Z_x de_{zx} + X_y de_{xy}\,. \tag{15}$$

Since the strain energy may be regarded as a single-valued function of the strains, it follows that

$$\frac{\partial X_x}{\partial e_{yy}} = \frac{\partial Y_y}{\partial e_{xx}};$$

and hence, from equations (11),

$$c_{12} = c_{21} .$$

Similarly, one finds that, in general,

$$c_{jk} = c_{kj} . \tag{16}$$

Upon forming the differential of $U - (X_x e_{xx} + Y_y e_{yy} + \ldots)$ in place of U, it follows that

$$s_{jk} = s_{kj} . \tag{17}$$

The matrices $\|s\|$ and $\|c\|$ are therefore symmetrical. These symmetry relations reduce the number of independent constants from thirty-six to twenty-one.

The numerical value of the elastic constants or coefficients depends upon the particular Cartesian coordinates to which they refer. It is sometimes desirable to express the elastic moduli with respect to one Cartesian system in terms of those referred to in another Cartesian system. This relation may readily be obtained by means of equations (5), (6), and (10), and of equations analogous to equations (9), which express the $X_x, \ldots,$ stress components in terms of the $X'_{x'}$ stress components. These latter equations may be derived in the same manner as were equations (9). They are

$$\left.\begin{aligned} X_x &= l_1^2 X'_{x'} + l_2^2 Y'_{y'} + l_3^2 Z'_{z'} + 2l_2 l_3 Y'_{z'} + 2l_3 l_1 Z'_{x'} + 2l_1 l_2 X'_{y'}, \text{ etc. }, \\ &\text{and} \\ X_y &= l_1 m_1 X'_{x'} + l_2 m_2 Y'_{y'} + l_3 m_3 Z'_{z'} + (l_2 m_3 + l_3 m_2)\, Y'_{z'} \\ &\quad + (l_3 m_1 + l_1 m_3)\, Z'_{x'} + (l_1 m_2 + l_2 m_1)\, X'_{y'} . \end{aligned}\right\} \tag{18}$$

Several examples will be given for crystals of cubic symmetry, in which, as is shown later, the elastic constants are specified by the following matrix:

$$\begin{Vmatrix} s_{11} & s_{12} & s_{12} & 0 & 0 & 0 \\ s_{12} & s_{11} & s_{12} & 0 & 0 & 0 \\ s_{12} & s_{12} & s_{11} & 0 & 0 & 0 \\ 0 & 0 & 0 & s_{44} & 0 & 0 \\ 0 & 0 & 0 & 0 & s_{44} & 0 \\ 0 & 0 & 0 & 0 & 0 & s_{44} \end{Vmatrix}$$

a) As a first example, we shall obtain the constant s'_{11} referred to the x'-, y'-, and z'-axes in terms of the constants referred to the x-, y-, and z-axes. According to equation (5), we multiply the first of equations (10) by l_1^2, the second by m_1^2, etc., and then form the sum, thereby obtaining e'_{xx}. We then transform the stresses in the right-hand side according to equations (18). The coefficient of $X'_{x'}$, defined as s'_{11}, is given by

$$s'_{11} = s_{11} - \{2(s_{11} - s_{12}) - s_{44}\} \cdot \{l_1^2 m_1^2 + m_1^2 n_1^2 + n_1^2 l_1^2\}. \tag{19}$$

b) As a second example, we shall find the elastic shear constant for the (110)[$\bar{1}$10] shear of a cubic crystal. In this case our table of direction cosines is as follows:

	x	y	z
x'	$\frac{1}{\sqrt{2}}$	$\frac{1}{\sqrt{2}}$	0
y'	$-\frac{1}{\sqrt{2}}$	$\frac{1}{\sqrt{2}}$	0
z'	0	0	1

Upon applying this table to equation (6), we obtain

$$e_{y'z'} = e_{zz} - e_{yy}.$$

Now, when we express the strains in terms of the stresses by means of the $\|s\|$ matrix for a cubic crystal and then express the stresses referred to xyz-axes into stresses referred to $x'y'z'$-axes by means of equations (18), we obtain

$$e_{y'z'} = 2(s_{11} - s_{12})\, Y'_{z'};$$

and hence

$$s'_{44} = 2(s_{11} - s_{12}). \tag{20}$$

c) As a third example, we shall find the shear constant corresponding to the (111)[$\bar{2}$11] shear of a cubic crystal. Our table of direction cosines now is:

	x	y	z
x' [11$\bar{2}$]	$\frac{1}{\sqrt{6}}$	$\frac{1}{\sqrt{6}}$	$-\frac{2}{\sqrt{6}}$
y' [111]	$\frac{1}{\sqrt{3}}$	$\frac{1}{\sqrt{3}}$	$\frac{1}{\sqrt{3}}$
z' [1$\bar{1}$0]	$\frac{1}{\sqrt{2}}$	$-\frac{1}{\sqrt{2}}$	0

Proceeding as before, we obtain

$$e_{x'y'} = \tfrac{1}{18}\{2\,(e_{xx} + e_{yy} - 2\,e_{zz}) + (2\,e_{xy} - e_{yz} - e_{zx})\}$$

$$= \tfrac{1}{3}\,(s + 2\,s')\,X'_{y'} + \frac{1}{\sqrt{18}}\,(s' - s)\,(Z'_{z'} - X'_{x'})\,,$$

where

$$s = s_{44}\,, \qquad s' = 2\,(s_{11} - s_{12})\,.$$

We therefore obtain

$$s'_{44} = \tfrac{1}{3}\{\,s_{44} + 4\,(s_{11} - s_{12})\,\}. \tag{21}$$

A review of the methods of determining the elastic constants has been given by Hearmon.[2]

D. EFFECT OF CRYSTAL SYMMETRY

As previously mentioned, the values of the elastic coefficients vary with the orientation of the coordinate axes with respect to the crystal. If one of the coordinate axes coincides with an axis of symmetry of the crystal, the relative orientations of the coordinate axes and the crystal are unchanged by an appropriate rotation of the coordinate axes about the axis of symmetry. The elastic coefficients must therefore also remain unchanged. This invariance is possible only if they satisfy certain conditions given in the accompanying table for simple symmetry operations.[3]

RESTRICTIONS PLACED UPON ELASTIC COEFFICIENTS BY SYMMETRY RELATIONS

$$\left\|\begin{array}{cccccc} c_{11} & c_{12} & c_{13} & 0 & 0 & c_{16} \\ c_{12} & c_{22} & c_{23} & 0 & 0 & c_{26} \\ c_{13} & c_{23} & c_{33} & 0 & 0 & c_{36} \\ 0 & 0 & 0 & c_{44} & c_{45} & 0 \\ 0 & 0 & 0 & c_{45} & c_{55} & 0 \\ c_{16} & c_{26} & c_{36} & 0 & 0 & c_{66} \end{array}\right\|$$

z-axis has a twofold symmetry, or reflection symmetry exists upon xy-plane

$$\left\|\begin{array}{cccccc} c_{11} & c_{12} & c_{13} & c_{14} & -c_{25} & 0 \\ c_{12} & c_{11} & c_{13} & -c_{14} & c_{25} & 0 \\ c_{13} & c_{13} & c_{33} & 0 & 0 & 0 \\ c_{14} & -c_{14} & 0 & c_{44} & 0 & c_{25} \\ -c_{25} & c_{25} & 0 & 0 & c_{44} & c_{14} \\ 0 & 0 & 0 & c_{25} & c_{14} & \frac{c_{11}-c_{12}}{2} \end{array}\right\|$$

z-axis has a threefold symmetry

$$\left\|\begin{array}{cccccc} c_{11} & c_{12} & c_{13} & 0 & 0 & c_{16} \\ c_{12} & c_{11} & c_{13} & 0 & 0 & -c_{16} \\ c_{13} & c_{13} & c_{33} & 0 & 0 & 0 \\ 0 & 0 & 0 & c_{44} & 0 & 0 \\ 0 & 0 & 0 & 0 & c_{44} & 0 \\ c_{16} & -c_{16} & 0 & 0 & 0 & c_{66} \end{array}\right\|$$

z-axis has a fourfold symmetry

$$\left\|\begin{array}{cccccc} c_{11} & c_{12} & c_{13} & 0 & 0 & 0 \\ c_{12} & c_{11} & c_{13} & 0 & 0 & 0 \\ c_{13} & c_{13} & c_{33} & 0 & 0 & 0 \\ 0 & 0 & 0 & c_{44} & 0 & 0 \\ 0 & 0 & 0 & 0 & c_{44} & 0 \\ 0 & 0 & 0 & 0 & 0 & \frac{c_{11}-c_{12}}{2} \end{array}\right\|$$

z-axis has a sixfold symmetry

2. R. F. S. Hearmon, *Rev. Mod. Phys.*, XVIII (1946), 409.

3. Voigt, *op. cit.*, pp. 581–89; A. E. H. Love, *Mathematical Theory of Elasticity* (Cambridge: Cambridge University Press, 1934), pp. 151–61; W. A. Wooster, *Crystal Physics* (Cambridge: Cambridge University Press, 1938), pp. 241–45.

The conditions satisfied by the elastic coefficients of a crystal having any type of symmetry may be obtained from this table. As an example, we shall consider a crystal with cubic symmetry. Here each of the three principal axes has a fourfold symmetry. Since the z-axis is no longer preferred, we obtain the following matrix for the elastic coefficients:

CRYSTAL OF CUBIC SYMMETRY

$$\left\| \begin{matrix} c_{11} & c_{12} & c_{12} & 0 & 0 & 0 \\ c_{12} & c_{11} & c_{12} & 0 & 0 & 0 \\ c_{12} & c_{12} & c_{11} & 0 & 0 & 0 \\ 0 & 0 & 0 & c_{44} & 0 & 0 \\ 0 & 0 & 0 & 0 & c_{44} & 0 \\ 0 & 0 & 0 & 0 & 0 & c_{44} \end{matrix} \right\|$$

Because of the dominant role of the cubic system in metals, it is desirable to have the relations between the elastic constants and coefficients in this system. It may readily be seen that the equations for the c's in terms of the s's are

$$c_{11} = \frac{s_{11} + s_{12}}{(s_{11} - s_{12})(s_{11} + 2s_{12})},$$

$$c_{12} = \frac{-s_{12}}{(s_{11} - s_{12})(s_{11} + 2s_{12})},$$

$$c_{44} = \frac{1}{s_{44}}.$$

From these equations the s's may be found in terms of the c's:

$$s_{11} = \frac{c_{11} + c_{12}}{(c_{11} - c_{12})(c_{11} + 2c_{12})},$$

$$s_{12} = \frac{-c_{12}}{(c_{11} - c_{12})(c_{11} + 2c_{12})},$$

$$s_{44} = \frac{1}{c_{44}}.$$

These relations may be expressed more succinctly in terms of the two shear constants, s_{44}, $2(s_{11} - s_{12})$, and the compressibility, $3(s_{11} + 2s_{12})$. These relations are

$$\left.\begin{aligned} s_{44} &= \frac{1}{c_{44}}, \\ s_{11} - s_{12} &= \frac{1}{c_{11} - c_{12}}, \\ s_{11} + 2s_{12} &= \frac{1}{c_{11} + 2c_{12}}. \end{aligned}\right\} \quad (22)$$

II

LOW-TEMPERATURE ELASTIC CONSTANTS IN CUBIC METALS

A. REVIEW OF DATA

THE elastic constants and coefficients of those cubic metals which have been studied are given in Table 1. A comparison of the values from different sources has been given by Hearmon.[1]

The coefficient c_{44} may be interpreted directly in terms of structure. It is a measure of the resistance to deformation with respect to a shearing stress applied across the (100) plane in the [010] direction. In a simple cubic structure of hard balls this resistance would be zero. On the other hand, the coefficients c_{11} and c_{12} have no such direct physical interpretation. It is therefore desirable to find linear combinations of these coefficients which have simple interpretations. Such linear combinations are $(c_{11} + 2c_{12})/3$ and $(c_{11} - c_{12})/2$. The first of these combinations may readily be seen to be the bulk modulus and is therefore a measure of the resistance to deformation with respect to a hydrostatic pressure. The second of these coefficients may readily be seen, from equations (20) and (22), to be the resistance to deformation by a shear stress applied across the (110) plane in the $[1\bar{1}0]$ direction. This resistance would be zero for a body-centered cubic lattice of hard balls. In a discussion of the physical interpretation of the elastic constants, attention should therefore be focused upon the following three coefficients:

$$K \equiv \frac{c_{11} + 2c_{12}}{3}, \qquad C \equiv c_{44}, \qquad C' \equiv \frac{c_{11} - c_{12}}{2}.$$

These coefficients are given in Table 3.

The relative magnitudes of K, C, and C' are frequently of interest and may be expressed in terms of dimensionless ratios, which are of interest per se. In an elastically isotropic body, C and C' must be identical. We shall therefore define

$$A \equiv \frac{C}{C'},$$

as the anisotropy factor. This ratio is likewise given in Table 3.

When a tensile stress, S, is applied along the x-axis, the resolved shear stress across the {100} planes in the corresponding <010> directions is

1. R. F. S. Hearmon, *Rev. Mod. Phys.*, Vol. XVIII (1946), Table V.

zero. The resulting deformation is therefore specified completely by the two constants, K and C'. In order to see clearly the role that these two constants play, we consider the stress system as composed of two parts, namely,

$$X_x = Y_y = Z_z = \frac{S}{3}$$

and

$$\frac{X_x}{2} = -Y_y = -Z_z = \frac{S}{3}.$$

TABLE 1

ELASTIC DATA FOR CUBIC METALS (ROOM TEMPERATURE)

Crystal Type	Metal	Constants in 10^{-12} cm²/dyne			Coeffs. in 10^{12} dyne/cm²			Ref.
		s_{11}	s_{12}	s_{44}	c_{11}	c_{12}	c_{44}	
b.c.c.......	Fe(α)	0.757	− 0.282	0.862	2.37	1.41	1.16	1
b.c.c.......	Na(210° K.)	53.5	−23.2	20.37	0.0555	0.0425	0.0491	3
b.c.c.......	K	82.3	−37.0	38.0	0.0459	0.0372	0.0263	7
b.c.c.......	W	0.257	− 0.073	0.660	5.01	1.98	1.51	1
b.c.c.......	β-brass	3.64	− 1.34	1.22	1.279	1.091	0.822	4
f.c.c.......	Al	1.59	− 0.58	3.52	1.08	0.622	0.284	1
f.c.c.......	Au	2.33	− 1.07	2.38	1.86	1.57	0.420	1
f.c.c.......	Ag	2.32	− 0.993	2.29	1.20	0.897	0.436	1
f.c.c.......	Cu	1.49	− 0.625	1.33	1.70	1.23	0.753	1
f.c.c.......	Pb	9.30	− 4.26	6.94	0.483	0.409	0.144	2
f.c.c.......	α-brass	1.94	− 0.835	1.39	1.47	1.11	0.72	1
f.c.c.......	Cu_3Au	1.34	− 0.565	1.51	2.25	1.73	0.663	5
	C(diamond)	0.14	− 0.043	0.23	9.2	3.9	4.3	6

NOTES FOR TABLE 1

1. E. Schmid and W. Boas, *Kristallplastizität* (Berlin: Springer, 1935), pp. 21, 200.
2. E. Goens and J. Weerts, *Phys. Zeitschr.*, XXXVII (1936), 321.
3. S. L. Quimby and S. Siegel, *Phys. Rev.*, LIV (1938), 293.
4. D. Lazarus, *Phys. Rev.*, LXXIV (1948), 1726.
5. S. Siegel, *Phys. Rev.*, LVII (1940), 537.
6. S. Bhagavantam and J. Bhimasenachar, *Nature* CLIV (1944), 546.
7. O. Bender, *Ann. d. Phys.*, XXXIV (1939), 359.

The first stress system results in the dilation specified by

$$e_{xx} = e_{yy} = e_{zz} = \frac{S}{9K}.$$

The second stress system results in the dilationless deformation,

$$\frac{e_{xx}}{2} = -e_{yy} = -e_{zz} = \frac{S}{6C'}.$$

The first stress system gives rise to a transverse expansion, the second to a transverse contraction. The ratio of the net transverse contraction to the longitudinal extension is therefore a measure of the relative value of C'

and of K. This ratio is called "Poisson's ratio" and is denoted by σ_{100}. From the above sets of deformations, it is readily seen that

$$\sigma_{100}=\frac{3K-2C'}{6K+2C'}.$$

Values of this ratio are given in Table 3. Poisson's ratio is positive or negative according to whether K is greater or less than $(2/3)C'$. The former is the case in all metals so far examined. A mineral has been studied in which the latter is the case, pyrite.[2] The elastic constants and the K, C, and C' of pyrite are given in Table 2.

When a tensile stress, S, is applied along the [111] axis, the resolved shear stress across the {101} planes along the corresponding $<\bar{1}01>$ directions is zero. The resulting deformation must therefore be determined

TABLE 2

EXAMPLE OF NEGATIVE POISSON'S RATIO

MINERAL	CONSTANTS ×10¹²*			COEFFS. ×10⁻¹²		
	s_{11}	s_{12}	s_{44}	K	C	C'
FeS_2.	0.289	0.044	0.948	0.89	1.05	2.04

* E. Schmid and W. Boas, *Kristallplastizität* (Berlin: Springer, 1935), p. 267.

solely by the two coefficients K and C. From equation (19) the tensile strain parallel to the [111] axis is

$$e_{11}=\left(\frac{1}{9K}+\frac{1}{3C}\right)S.$$

Since the total dilation is $(1/3K)S$, the strain transverse to the [111] axis is given by

$$e_{11}+2e_{\perp}=\frac{1}{3K}S.$$

From these two equations one finds, as the ratio of the transverse contraction to the tensile extension,

$$\sigma_{111}=\frac{3K-2C}{6K+2C}. \quad (23)$$

The Poisson ratio, σ_{111}, may therefore be regarded as a measure of the relative values of K and C. This ratio is given in Table 3. According to

2. The Poisson ratio of sodium chlorate was reported by Voigt as being negative but has recently been found to be positive (see W. P. Mason, *Phys. Rev.*, LXX [1946], 529; and S. Bhagavantam and D. Suryanarayan, *Phys. Rev.*, LXXI [1947], 553).

whether K is greater or less than $(2/3)C$, Poisson's ratio, σ_{111}, is positive or negative. From Table 3 it may be seen that β-brass is the only crystal as yet examined for which this ratio is negative, i.e., for which a specimen subjected to a tensile stress along a <111> axis extends in the transverse direction.

B. RELATION OF ELASTIC CONSTANTS TO INTERATOMIC FORCES AND TO STRUCTURE

In his classic analysis of the elasticity of crystals,[3] Cauchy deduced certain relations between the coefficients of elasticity. In this proof Cauchy assumed that all forces were central, i.e., acted along lines joining the centers of the atoms. Cauchy's proof was not entirely satisfactory, and many alternative derivations have been presented, of which the best known is due to Voigt.[4] The Cauchy relations, first explicitly stated by Voigt, are

TABLE 3

ELASTIC BEHAVIOR OF CUBIC METALS

(Derived from Table 1)

Metal	$K\times10^{-12}$	$C\times10^{-12}$	$C'\times10^{-12}$	A	σ_{100}	σ_{111}
Fe(α)	1.73	1.16	0.480	2.4	0.37	+0.24
Na	0.0468	0.0491	0.0065	7.5	.43	+ .11
K	0.040	0.0263	0.0044	6.3	.45	+ .23
W	3.00	1.51	1.51	1.00	.28	+ .29
β-brass	1.15	0.822	0.093	8.8	.46	+ .21
Al	0.77	0.284	0.230	1.23	.36	+ .34
Au	1.67	0.420	0.147	3.9	.46	+ .39
Ag	1.00	0.436	0.151	2.9	.46	+ .23
Cu	1.39	0.753	0.237	3.3	.42	+ .27
Pb	0.43	0.144	0.037	3.9	.46	+ .35
α-brass	1.23	0.72	0.180	4.0	.43	+ .26
Cu_3Au	1.90	0.663	0.180	3.7	.34	+ .45
C(diamond)	5.7	4.3	2.6	1.6	0.30	+0.20

$$c_{44} = c_{23}\,, \qquad c_{55} = c_{31}\,, \qquad c_{66} = c_{12}\,;$$

$$c_{56} = c_{14}\,, \qquad c_{64} = c_{25}\,, \qquad c_{45} = c_{36}\,.$$

In the case of a crystal with cubic symmetry, these relations reduce to

$$c_{44} = c_{12}\,.$$

Voigt's derivation has recently been shown by Epstein[5] to be not generally valid. From Epstein's analysis one may deduce,[6] however, as had already

3. A. L. Cauchy, "De la pression de tension dans un système de points matériels," *Exercices de mathématique* (1828).

4. W. Voigt, *Lehrbuch der Kristallphysik* (Berlin: Teubner, 1910), p. 608.

5. P. S. Epstein, *Phys. Rev.*, LXX (1946), 915.

6. C. Zener, *Phys. Rev.*, LXXI (1947), 323.

been shown by Love,[7] that the Cauchy relations are valid, provided that one additional assumption is made regarding the structure and provided that the crystal is initially under no stress. This additional assumption is that every atom is at a center of symmetry. Only when each atom is at a center of symmetry is the displacement of each atom completely specified by the macroscopic strain. This symmetry condition is satisfied by many ionic and metal crystals, namely, by simple cubic, body-centered cubic (b.c.c.), and face-centered cubic (f.c.c.) structures.

In ionic lattices the dominant forces arise from the electrostatic interaction of the ions. It is therefore to be expected that in those ionic crystals of appropriate symmetry the Cauchy relations will be approximately satisfied. From Table 4 this is seen to be, in fact, the case.

TABLE 4

TEST OF CAUCHY RELATIONS FOR IONIC LATTICES
(Room Temperature)

	Elastic Coefficients in Units of 10^{12} Dynes/Cm^2	
	c_{12}	c_{44}
LiF*	0.404	0.554
NaCl*,†	.123	.126
KBr*,†	.054	.051
KCl†	0.060	0.063

* H. B. Huntington, *Phys. Rev.*, LXXII (1947), 321.
† J. K. Galt, *Phys. Rev.*, LXXIII (1948), 1460.

Two dominant factors in metallic binding[8] are the distortion of the wave functions of the outermost electrons, giving rise to attraction, and the kinetic energy of these electrons, which increases with decreasing volume, giving rise to repulsion. Neither of these two types of energy may be regarded as resulting from forces acting along lines joining atoms, i.e., from central forces. One therefore anticipates that in metals the Cauchy relations will not be even approximately valid. This is indeed the case, as may be seen by reference to Table 1 (p. 17).

An understanding of the characteristically metallic contributions to the elastic coefficients may be best gained through a study of sodium, the metal whose properties conform most closely to those predicted from quantum-mechanical computations based upon simple assumptions. If we neglect the overlapping of the inner cores and the van der Waals in-

7. A. E. H. Love, *Mathematical Theory of Elasticity* (Cambridge: Cambridge University Press, 1934), Appen. B.

8. For discussions of metallic binding see F. Seitz, *Modern Theory of Solids* (New York: McGraw-Hill Book Co., 1940), chap. x; and N. Mott and H. Jones, *Properties of Metals and Alloys* (London: Oxford University Press, 1936), chap. iv.

teraction between the inner cores, an approximation which introduces only a slight error in the case of the alkali metals, we obtain for the energy per atom[9]

$$E = \phi(v) - X . \qquad (24)$$

In this equation ϕ is a function only of the atomic volume, v, and is independent of shear distortions, while the function X is primarily a function of shear distortions, being identically zero in the absence of such distortion. The binding energy and the bulk modulus, K, are thus completely determined by the function $\phi(v)$. This function may be found only through a detailed computation of the variation of the wave function with the lattice-spacing. Such a computation would have to be repeated for each metal studied. On the other hand, the shear coefficients are completely determined by X alone. This function arises from the quadrupole-quadrupole

TABLE 5

CONTRIBUTION OF X TO SHEAR COEFFICIENTS OF MONOVALENT METALS
(In Units of 10^{12} Ergs/Cm^3)

	f.c.c.	b.c.c.
c_{44}*.	$40.5a^{-4}$	$16.8a^{-4}$
$(c_{11}-c_{12})/2$.	$4.52a^{-4}$	$2.26a^{-4}$

* a: Lattice constant in units of 10^{-8} cm.

electrostatic interaction associated with the distortions of the individual atoms. In all monovalent metals of the same lattice type, X is the same function of shear strains and of lattice parameters. It thus need be computed only once for all monovalent metals of the same lattice type. Fuchs has performed these computations and has obtained the results given in Table 5.

A comparison of the computed and experimental elastic coefficients for sodium are given in Table 6. The close agreement of theoretical and experimental values of the elastic coefficients is evidence of the essential correctness of the assumptions upon which the computations were based.

In metals other than the alkalis the noncoulombic interaction of the ions gives an important contribution to the elastic coefficients. This noncoulombic interaction is of the Cauchy type, acting along a line joining the centers of the two interacting ions. The energy per atom due to this interaction is

$$E = \tfrac{1}{2}\Sigma W(r) ,$$

9. K. Fuchs, *Proc. Roy. Soc., London*, CLI (1935), 585; CLIII (1936), 622; CLVII (1936), 444.

where $W(r)$ is the interaction energy of a single pair of ions at a distance r apart and the summation is over all the neighbors of a particular reference atom. The contribution to the elastic coefficients, K, $(c_{11} - c_{12})/2$, and c_{44}, will then be $N\partial^2E/\partial s^2$, where N is the number of atoms per unit volume and s is defined by

$$\delta r^3 = r^3 s ,$$

$$\delta (x, y, z) = \left\{\frac{x+y}{2} s , -\frac{x+y}{2} s , 0\right\} ,$$

$$\delta (x, y, z) = (y s, 0, 0) ,$$

respectively. It may readily be seen that the contribution to the two shear coefficients is

$$\left\{\begin{matrix}\Delta C \\ \Delta C'\end{matrix}\right\} = \tfrac{1}{2} N \Sigma \left\{\left(\frac{\partial r}{\partial s}\right)^2 W'' + \frac{\partial^2 r}{\partial s^2} W'\right\} . \tag{25}$$

TABLE 6

ELASTIC COEFFICIENTS OF SODIUM
(In Units of 10^{12} Dynes/Cm^2)

	Theoretical*	Experimental† (80° K.)
K	0.062	0.051
c_{44}	.058	.059
$(c_{11}-c_{12})/2$	0.0070	0.0072

* K. Fuchs, *Proc. Roy. Soc., London,* CLI (1935), 585; CLIII (1936), 622; CLVII (1936), 444.

† S. L. Quimby and S. Siegel, *Phys. Rev.*, LIV (1938), 293.

The noncoulombic ionic interaction is essentially short ranged. Only a slight error is therefore introduced by regarding the summation in equation (25) as extending over only the nearest neighbors. We then see that, if the crystal type and shear are such that $(\partial r/\partial s)$ is exactly zero for all closest neighbors, equation (25) becomes

$$\left\{\begin{matrix}\Delta C \\ \Delta C'\end{matrix}\right\} = \tfrac{1}{2} N \Sigma \frac{\partial^2 r}{\partial s^2} W' . \tag{26}$$

Such is the case, for example, for a (100)[010] shear in a simple cubic structure, and for a (110) [$\bar{1}$10] shear in a b.c.c. lattice. In these cases ΔC or $\Delta C'$ have the same sign as W', being negative when W corresponds to a repulsive force. It may be noted that it is just these cases for which a packing of hard balls would offer no shear resistance.

Several interesting conclusions may be deduced from equation (26). If the ions consist of closed shells, the interaction will certainly be repulsive and hence the contribution to ΔC or to $\Delta C'$ will be negative. If this negative contribution is larger than the positive contribution due to X, the lattice will be mechanically unstable. One thereby deduces that metals such as copper would be mechanically unstable as b.c.c. structures. As will be pointed out in chapter iv, such instability would cause the lattice automatically to shear into a f.c.c. lattice. Such considerations lead one to expect to find b.c.c. structures only when the ions contain incomplete inner shells. Aside from the alkalis, where the noncoulombic interaction is negligible, b.c.c. structures are found in V, Cr, Fe, Cb, Mo, Ta, W, Eu, which satisfy this requirement, and in Ba and Tl, which form exceptions. In alloys further exceptions are found in the β-brass type b.c.c. phases. In these phases, however, the concentration range of the disordered b.c.c. phase rapidly narrows as the temperature is lowered, the phase disappearing if the critical temperature for ordering is not reached in time. It thus appears that these phases are almost, but not quite, mechanically unstable. This conclusion is supported by the observation that in β-brass, the only alloy of this type whose elastic coefficients have been measured, the $(c_{11} - c_{12})/2$ coefficients is extremely small, being only one-ninth that of c_{44}. The positive temperature coefficient of $(c_{11} - c_{12})/2$ in β-brass has likewise been interpreted[10] in terms of equation (26). As the lattice expands with a rise in temperature, the negative part of this elastic coefficient due to W' decreases much faster than does the positive part, thereby resulting in a net increase. An interpretation is given in chapter iv of f.c.c. $\rightarrow$ b.c.c. transitions based upon the low value of the $(c_{11} - c_{12})/2$ shear coefficient of b.c.c. structures.

10. C. Zener, *Phys. Rev.*, LXXI (1947), 846.

III

TEMPERATURE DEPENDENCE OF ELASTIC CONSTANTS

THE elastic moduli of crystalline materials normally decrease with increasing temperature, although a few exceptions are known. At least part of this normal decrease is frequently attributable to some sort of relaxation. An example is given in Figure 4 (after Kê).[1] In this case the relaxa-

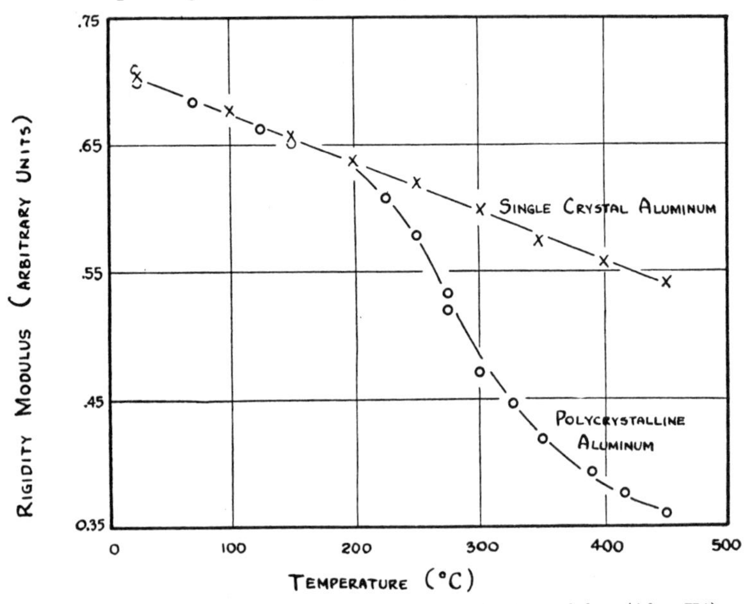

Fig. 4.—Illustration of effect of relaxation upon elastic modulus. (After Kê)

tion of shear stress by viscous slip across grain boundaries decreases the observed modulus by 33 per cent. The effect of relaxations upon the elastic moduli is discussed in detail in Part Two of this monograph. We are at present concerned with the effect of temperature in the absence of any relaxation. In such cases the normal variation of the elastic coefficients with temperature is linear over a wide temperature range. This linear variation holds from well below the Debye characteristic temperatures to near the melting temperatures. Examples are illustrated in Figures 5 (after Quimby and Siegel)[2] and 6 (after Rose,[3] Durand,[4]

1. T. S. Kê, *Phys. Rev.*, LXXI (1947), 533.
2. S. L. Quimby and S. Siegel, *Phys. Rev.*, LIV (1938), 293.
3. F. Rose, *Phys. Rev.*, XLIX (1936), 50.
4. M. A. Durand, *Phys. Rev.*, L (1936), 449.

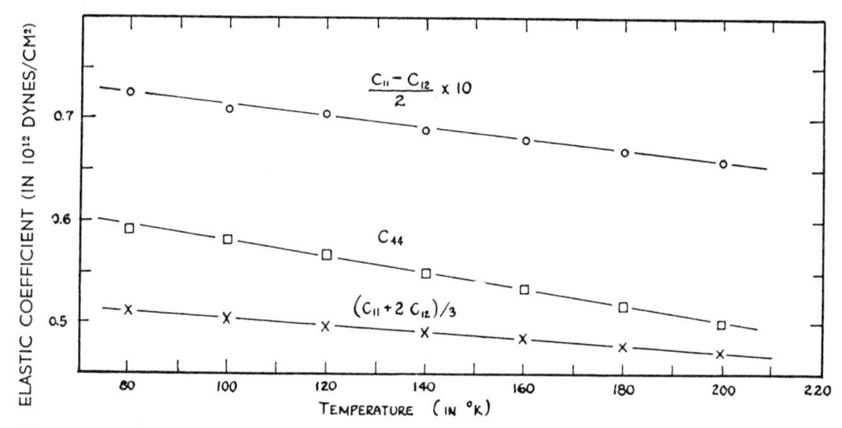

FIG. 5.—Temperature variation of elastic coefficients of sodium. (After Quimby and Siegel)

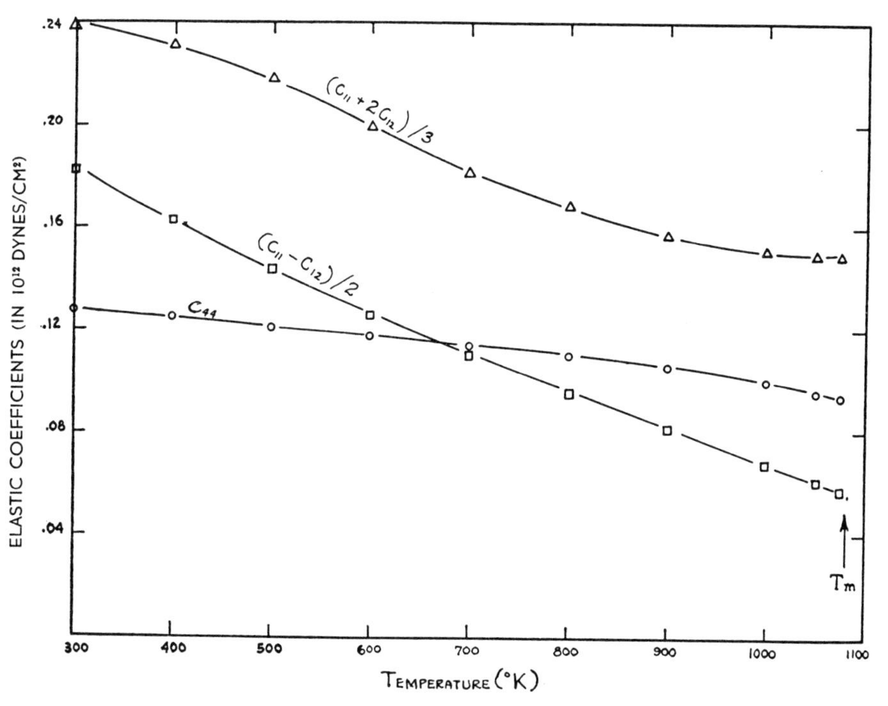

FIG. 6.—Temperature variation of elastic coefficients of NaCl. (After Rose, Durand, Hunter, and Siegel.)

Hunter and Siegel[5]). The normal linear decrease of the elastic coefficients with temperature indicates that the elastic coefficients decrease in essentially a linear manner with thermal energy.

A discussion of the temperature dependence of the elastic coefficients would not be complete without an analysis of the difference between the coefficients measured under isothermal and adiabatic conditions. Of the three coefficients of a cubic crystal—c_{11}, c_{12}, and c_{44}—the last is the same under both conditions, i.e., the application of a (100)[010] shear does not lead to a change in temperature. As we have found in previous discussions, here also a simplification is introduced by an analysis of $(c_{11} + 2c_{12})/3$, $(c_{11} - c_{12})/2$, and c_{44} rather than of c_{11}, c_{12}, and c_{44}. Since $(c_{11} - c_{12})/2$ as well as c_{44} is identical when measured isothermally or adiabatically, as may readily be deduced from symmetry conditions, a discussion of the difference between the isothermal and the adiabatic coefficients reduces to an analysis of only the bulk modulus, K.

The physical principles involved in the temperature variation of the elastic coefficients may most readily be understood from an analysis of a single particle in a potential valley. We therefore start our analysis with such a simple system.

A. SINGLE-PARTICLE MODEL

The first example we shall study is that of a single particle in a symmetrical anharmonic potential, as illustrated in Figure 7. Zero motion will correspond to 0° K., vibrational energy corresponding to thermal energy. At 0° K. the force constant, C, is given by

$$C = V''(0) .$$

Let us now consider that the particle has thermal energy, i.e., vibrational energy, which, when no external force is applied, has the value E_0. Let us now consider that an external force, f, is applied, the time during which this force builds up to its final value being very long compared with the period of vibration of the particle. As is well known, the vibrational energy no longer remains constant under these conditions but follows the equation

$$E(\bar{x}) \backsim \omega(\bar{x}) ,$$

where $\bar{x}$ is the mean position of the particle and ω is the angular frequency. This equation follows directly upon substitution of the approximate solution,

$$X(t) = A(t) \cos \int^t \omega(t)\, dt$$

5. L. Hunter and S. Siegel, *Phys. Rev.*, LXI (1942), 84.

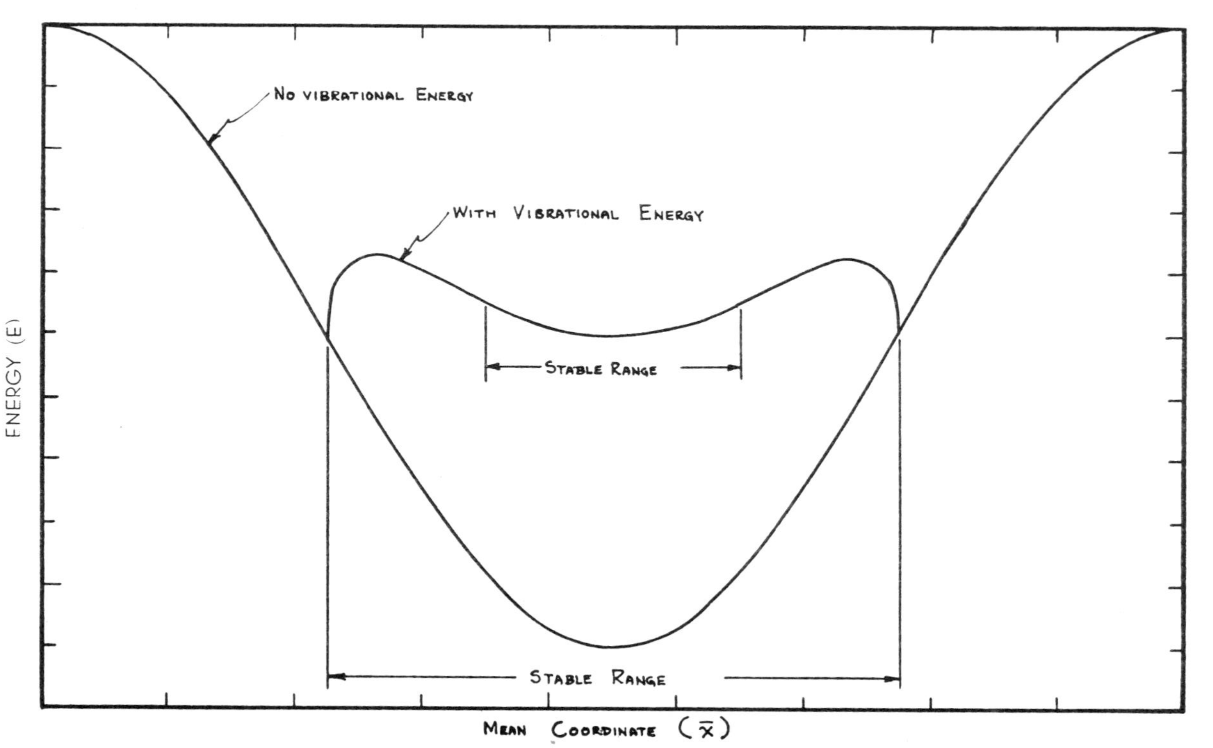

Fig. 7.—Illustration of the influence of thermal energy upon adiabatic elastic coefficients

into the differential equation for $X(t)$, neglecting A''. The total energy, E, may therefore be expressed as a function of $\bar{x}$ as

$$E(\bar{x}) = V(\bar{x}) + \frac{E_0 \omega(\bar{x})}{\omega_0}, \tag{27}$$

where E_0 and ω_0 refer to the vibrational energy and angular frequency when no force is present, respectively. Since ω varies as $(V'')^{1/2}$, the above equation becomes

$$E(\bar{x}) = V(\bar{x}) + E_0 \left\{ \frac{V''(\bar{x})}{V''(0)} \right\}^{1/2}. \tag{28}$$

A typical example of the influence of vibrational energy upon $E(\bar{x})$ is given in Figure 7. In this example we have taken

$$V(x) = V_0 \left(1 - \cos \frac{2\pi x}{\lambda} \right) \tag{29}$$

and

$$E_0 = V_0 .$$

In this example the thermal energy reduces the adiabatic elastic constant $E''(0)$ by 50 per cent; reduces the range of stability, i.e., the range in which $E''(\bar{x})$ is positive, by 60 per cent; and reduces the maximum elastic energy by a factor of nearly 0.1. In the more general case, in which $V(x)$ is given by equation (29) and E_0 has an arbitrary value, one finds for the adiabatic coefficient

$$C = V''(0) \left\{ 1 - \frac{1}{2} \frac{E_0}{V_0} \right\}. \tag{30}$$

The symmetrical potential of Figure 7 might be regarded as referring to the potential associated with a (100)[010] or a (110)[$\bar{1}$10] shear, i.e., with c_{44} or $(c_{11} - c_{12})/2$. On the other hand, a potential associated with dilation, i.e., with the bulk modulus K, must be unsymmetrical. If the origin of coordinates is chosen to be at the minimum of the potential, then $V'(0)$ will be zero, but the higher odd derivatives, such as $V'''(0)$, will, in general, not vanish. The adiabatic coefficient for an unsymmetrical potential will be given by the second derivative of $E(\bar{x})$ in equation (28), $\bar{x}$ having that value acquired from thermal expansion. It may be obtained either by forming the average,

$$\bar{x} = \frac{\int_{-\infty}^{\infty} x\, e^{-V(x)/kT}\, dx}{\int_{-\infty}^{\infty} e^{-V(x)/kT}\, dx},$$

or by minimizing the free energy,

$$F = V(\bar{x}) - \frac{kT \ln kT}{h\nu(\bar{x})}. \tag{31}$$

Upon neglecting terms containing kT to a power higher than the first, we obtain from these two equations

$$\bar{x} = -\frac{1}{2}\frac{V'''}{V''^2}kT \tag{32}$$

and

$$\bar{x} = -\frac{\nu'}{\nu}\frac{kT}{\nu''}, \tag{33}$$

respectively, the derivatives being taken at $\bar{x} = 0$. These two expressions are, of course, equivalent.

The adiabatic elastic coefficient, C_S, is obtained by taking the second derivatives of $E(\bar{x})$ in equation (27). We obtain

$$C_S = V_0'' - \frac{1}{2}\left(\frac{V'''}{V''}\right)^2 kT + \frac{\nu''}{\nu}kT. \tag{34}$$

On the other hand, the isothermal elastic coefficient, C_T, is obtained by taking the second derivative of the free energy, as given by equation (31). We find

$$C_S - C_T = \left(\frac{\nu'}{\nu}\right)^2 kT$$

$$= \frac{1}{4}\left(\frac{V'''}{V''}\right)^2 kT.$$

B. REAL CRYSTALS

It is customary to assume that the normal modes of vibration of a crystal interfere with one another only through thermal expansion. To this approximation one may write, for temperatures above the characteristic temperatures,

$$E = V + \sum_j n_j h\nu_j, \tag{35}$$

$$F = V - kT\sum_j \ln\frac{kT}{h\nu_j}. \tag{36}$$

In these equations the summation extends over all normal modes of vibration, and the potential energy, V, and the frequencies are functions of the macroscopic distortions but are independent of the n_j's and of T except implicitly through the dilation θ. Although the potential V is a symmetrical function of the shear strains (100)[010] and (110)[$\bar{1}$10], nonetheless a rise in temperature will influence the elastic coefficients c_{44} and

$(c_{11} - c_{12})/2$ both through the influence of thermal energy per se and through thermal expansion. Thus, if s denotes the (100)[010] shear strain, the effect of thermal expansion will come about because of the nonvanishing of $(\partial^3 V/\partial\theta\partial s^2)$.

The thermal dilation, θ, is found by taking that value of θ which minimizes the free energy, F. Upon letting ν denote the geometrical mean of all the frequencies, we obtain from equation 36

$$\theta = -\frac{3NkT}{K}\frac{\partial ln\nu}{\partial\theta}. \qquad (37)$$

Here N is the number of atoms per unit volume, and K is the bulk modulus. The derivative $-\partial ln\nu/\partial\theta$ has been given the symbol γ by Grüneissen,[6]

$$\gamma \equiv -\frac{\partial ln\nu}{\partial\theta}.$$

Since ν is defined as the geometrical mean of all the frequencies, γ depends upon the volume variation of all the elastic coefficients, e.g., upon K, c_{44}, and $(c_{11} - c_{12})/2$. Now in b.c.c. lattices the volume variation of $(c_{11} - c_{12})/2$ is very small. In such lattices $(c_{11} - c_{12})/2$ consists of two terms.[7] One term, arising from the electrostatic interaction of conduction electrons and ions, varies as $1/a^4$, where a is the lattice constant. The contribution of the second term, arising from the exchange interaction of adjacent ions, increases with increasing volume. Thus a frequency ν_1 which involves solely the elastic coefficient $(c_{11} - c_{12})/2$ in a body-centered metal will vary as

$$-\frac{\partial ln\nu_1}{d\theta} < \tfrac{1}{6}.$$

As in the case of β-brass, such frequencies may actually increase with increasing volume.[8] One therefore anticipates that γ will be smaller for b.c.c. metals than for f.c.c. metals. Druyvesteyn[9] has already pointed out that such is the case, with the single exception of Ca. A comparison of the two types of cubic metals is given in Table 7.

If we now let s refer to either of the shear distortions—(100)[010] or (110)[$\bar{1}$10]—or to a dilation, we obtain the corresponding adiabatic modulus by twice differentiating E in equation (35) with respect to s, obtaining

$$C_S = V'' + 3NK^{-1}\gamma\frac{\partial^3 V}{\partial\theta\partial s^2}kT + \frac{\nu''}{\nu}kT. \qquad (38)$$

6. E. Grüneissen, *Ann. d. Phys.*, XXXIX (1912), 257.

7. N. Mott and H. Jones, *Properties of Metals and Alloys* (London: Oxford University Press, 1936), p. 148.

8. C. Zener, *Phys. Rev.*, LXXI (1947), 846.

9. M. J. Druyvesteyn, *Philips Res. Repts.*, I (1946), 77.

Here the first term refers to the elastic modulus in the absence of any vibrational energy. The second term is the change in the modulus introduced by thermal expansion. Finally, the third term is the direct effect of thermal energy, discussed at length above.

As already mentioned, the second and third terms in equation (38) are due, directly or indirectly, to thermal vibrational energy. On the other hand, special effects may cause the first term likewise to vary with temperature. Thus in β-brass the disordering effect of temperature is associated with a marked decrease in all the elastic coefficients.[10]

TABLE 7

COMPARISON OF GRÜNEISSEN'S CONSTANT FOR b.c.c. AND FOR f.c.c. METALS

(After Druyvesteyn* and Grüneisen†)

f.c.c. Metal	γ	b.c.c. Metal	γ
Au	3.03	W	1.62
Pb	2.73	Fe	1.60
Ir	2.54	Nb	1.59
Pt	2.54	Rb	1.48
Ag	2.40	Ti‡	1.40
Pd	2.23	K	1.34
Rh	2.20	V	1.30
Al	2.17	Cs	1.29
Cu	1.96	Na	1.25
Ni	1.88	Li	1.17
Co	1.87	Zr‡	0.96
Ca	1.14	Cr	0.94
Ta	1.75	Ba	0.86
Mo	1.66		

* M. J. Druyvesteyn, *Philips Res. Repts.*, I (1946), 77.
† E. Grüneissen, *Handb. d. Physik*, Vol. IX (1926).
‡ h.c.p. at room temperature, h.c.c. at high temperatures.

For a detailed account of the theory of the temperature dependence of the elastic constants and of the possible relation of this variation to the phenomenon of melting, the reader is referred to the work of Herzfeld and Goeppert-Mayer,[11] and Born and his collaborators.[12]

10. W. A. Good, *Phys. Rev.*, LX (1941), 605.
11. K. Herzfeld and M. Goeppert-Mayer, *Phys. Rev.*, XLVI (1934), 995.
12. M. Born *et al.*, *Jour. Chem. Phys.*, VII (1939), 591; *Proc. Cambridge Phil. Soc.*, XXXVI (1940), 160, 173, 454, 466; XXXVII (1941), 34, 177; XXXVIII (1942), 61, 67, and 82; XXXIX (1943), 101, 104, 113; XL (1944), 151; R. Furth, *Proc. Roy. Soc.*, *London*, CLXXXIII (1944), 87.

IV

MICROELASTICITY

THE term "elasticity" as commonly used relates to the elastic response of a macrospecimen to applied stresses. In macrospecimens the strain never exceeds a value appreciably higher than 1 per cent before either plastic deformation or fracture ensues. The subject of elasticity is therefore confined to small strains; and, because of this restriction, Hooke's law is generally applicable, namely, stress and strain are proportional. On the other hand, it is commonly assumed that crystallographically perfect elementary regions, of, say, a hundred unit cells to an edge, will not suffer plastic deformation or fracture until the stress exceeds the theoretical limit. In fact, plastic deformation will occur simply by the elementary region's passing from one configuration of minimum energy to another by a homogeneous strain. The term "microelasticity" will refer to the relation between energy and strain of such an elementary region.

The microelastic strain energy is a periodic function of the shear strain. As an example we shall consider the case of simple shear across the (111) plane of a face-centered cubic lattice. If the shear is in the $[\bar{2}11]$ direction and is of sufficient magnitude, a twinned structure will result, as illustrated in Figure 8. In this figure the atoms in one (111) plane are denoted by crosses. The atoms in the second plane are denoted by circles. During the twinning deformation these atoms move from the positions denoted by open circles to the positions denoted by filled circles. Similarly, the atoms in the third plane move with respect to the atoms in the second plane, as these move with respect to those in the first plane, et cetera. During this twinning the elementary region passes from one minimum configuration to another. The essential periodicity of the strain energy with respect to such deformation is illustrated in Figure 9. In this figure the contour lines are given for the energy of a microelement at 0° K. The coordinates may be regarded as the coordinates of a particular atom in the second plane, or perhaps simply as the two components of shear strain across the (111) plane.

In Figure 9 the A-type positions correspond to a f.c.c. lattice and are the positions of minimum energy. Twinning corresponds to passing from one A position to an adjacent A position. Continued twinning, alternately in the $[\bar{2}11]$ and $[\bar{1}2\bar{1}]$ directions, will lead to a continuing deformation in the [$\bar{1}$10] direction, as illustrated in Figure 10. Thus a continued plastic deformation may be represented in Figure 9 by a zigzagging along the valleys between the maxima of type C. Along this path subsidiary minima may be present in positions of type B. These subsidiary minima are separated from the primary minima by divides in positions of type D. From the above description of plastic deformation it is evident that the resistance of an unhardened metal to plastic deformation is determined by the variation of the energy in going from one A position over a divide D, through a subsidiary minimum B, and over a second divide into an adjacent A position.

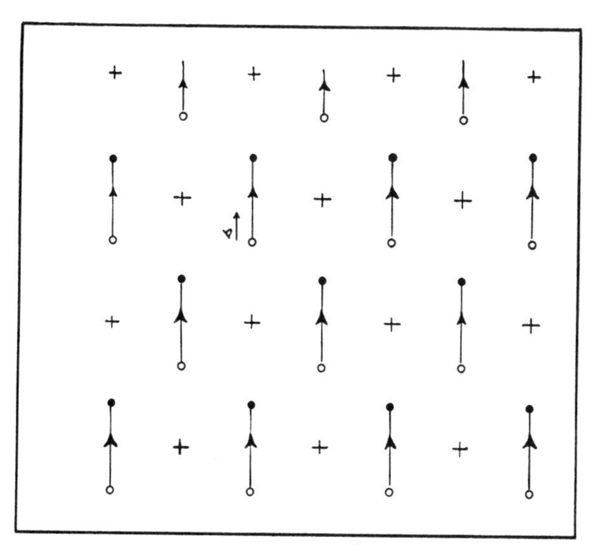

FIG. 8.—Atomic motions during twinning in a f.c.c. lattice. Plane of atoms is (111), direction of motion is [$\bar{2}$11].

We shall not further discuss the interesting subject of the relation of microelasticity to plastic deformation but shall restrict ourselves to a discussion of the variation of the microelastic energy with shear strain.

A microelastic energy plot may be constructed as in Figure 9 according to two schemes. According to one scheme, one takes as independent variables all the six strain components. A diagram such as Figure 9 then corresponds to all the strain components held constant except the two shear strains across the (111) plane. According to the second scheme, the only strain components taken as independent variables are the two shear strains across the (111) plane, while all the stress components are held constant except the two corresponding to these two shear strain components. Thus, in passing from configuration A to B, the distance between the (111) planes will remain invariant according to the first scheme but will change so as to maintain a minimum energy according to the second scheme. It is believed that this second scheme of representation is more useful, and it will therefore be adopted in the following discussion.

FIG. 9.—Example of microelastic contour lines

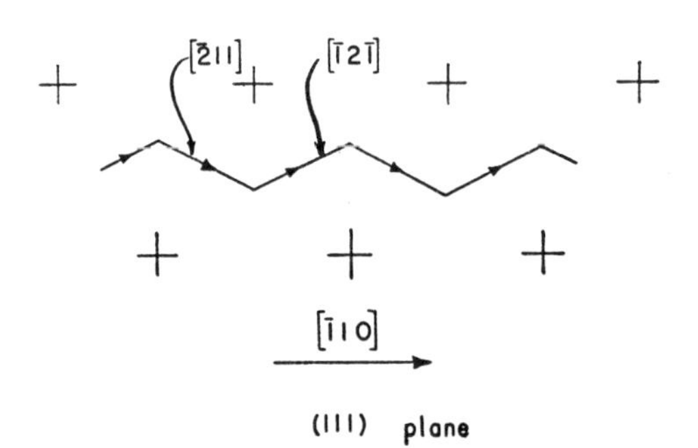

FIG. 10.—Illustration of how an unlimited amount of shear strain may occur in a f.c.c. structure through a succession of twin deformations.

Suppose that a shear twin proceeds only halfway, so that the atoms in the second plane of Figure 8 lie midway between the open and the filled circles. The resulting configuration is very close to that of a b.c.c. lattice, with its (110) plane and $[1\bar{1}0]$ direction coinciding with the (111) plane and $[\bar{2}11]$ direction of the original f.c.c. lattice, as may be seen by reference to Figure 11. If the b.c.c. lattice is mechanically stable, the lattice will automatically acquire the b.c.c. structure, so that configuration B in Figure 9 corresponds to a b.c.c. lattice. The free energy as a function of the twinning shear coordinate may therefore be represented as in Figure 12.

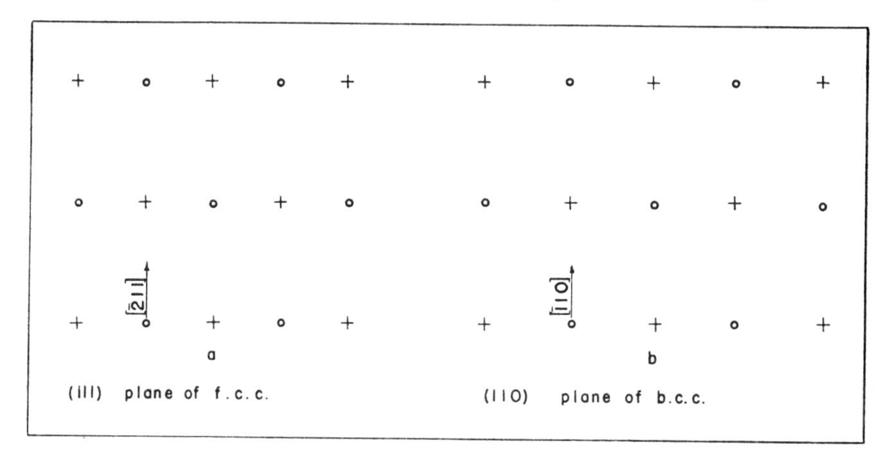

FIG. 11.—Illustration of readjustment which causes a f.c.c. lattice to pass through a b.c.c structure during a twin shear strain.

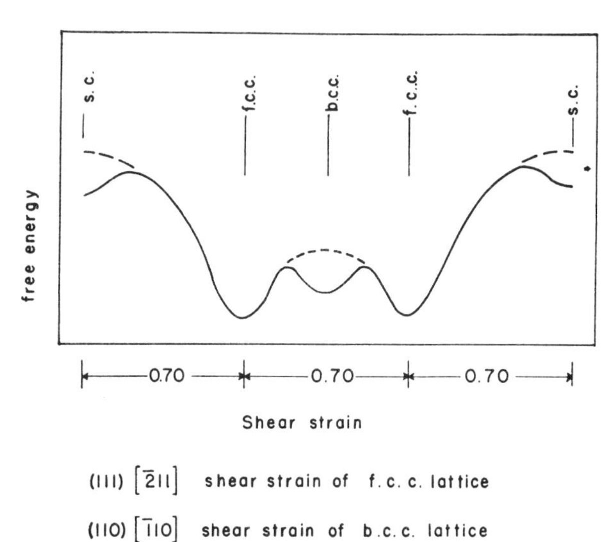

FIG. 12.—Illustration of microelastic strain energy

The inner contour lines around configurations of type A in Figure 9 are circles, corresponding to the fact that the elastic constant associated with a shear across a (111) plane is independent of the direction of this shear. Its value is given by

$$s_{(111)} = \tfrac{1}{3}\, s_{44} + \tfrac{2}{3} \cdot 2\,(s_{11} - s_{12})\,. \tag{39}$$

On the other hand, the inner contours around configurations of type B are ellipses, corresponding to the fact that the shear elastic constant across a (110) plane is dependent upon the direction of the shear. Thus

$$s_{(110)[\bar{1}10]} = 2\,(s_{11} - s_{12})\,, \tag{40}$$

while

$$s_{(110)[001]} = s_{44}\,. \tag{41}$$

In metals the first constant is always considerably larger than the second, the ratio of the two being as high as 18 in β-brass. The inner

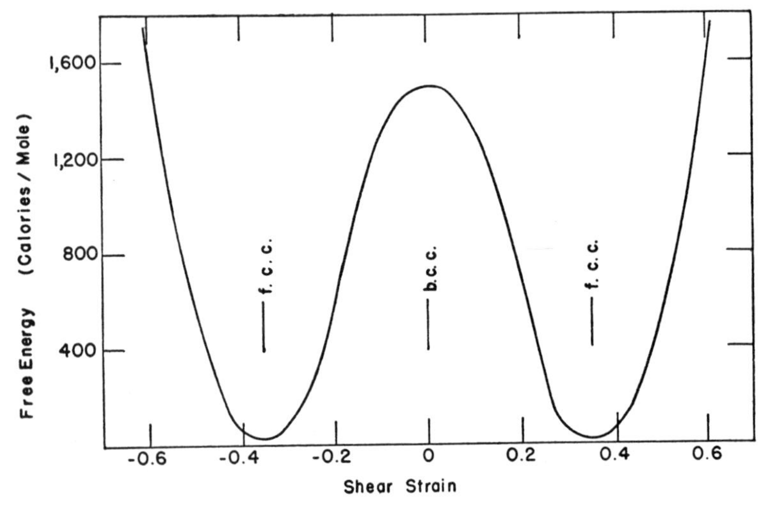

FIG. 13.—$F-s$ curve for copper, illustrating instability of b.c.c. structure

contours are therefore elongated along the axis joining the adjacent A-type positions.

From equations (39) and (40) an estimate of an energy curve such as that in Figure 9 may be obtained from a knowledge of the elastic constants of the f.c.c. and of the b.c.c. structures. An example will be given for copper. Here we know the elastic shear coefficients of the f.c.c. structure, namely, c_{44} and $(c_{11} - c_{12})/2$ are 0.75×10^{12} and 0.24×10^{12} dynes/cm^2, respectively. From the computations of Fuchs we may estimate the elastic coefficients for the b.c.c. structure. We obtain for $(c_{11} - c_{12})/2$ the value -0.02×10^{12} dynes/cm^2. An energy curve satisfying the appropriate curvatures is given as Figure 13. The height of the potential barrier is less than RT_m, where T_m is the melting temperature. This relatively small height of the potential barrier separating two twin configurations suggests that the structure of the crystalline and of the liquid state may be essentially the same, save that the lattice contains many local fluctuations of structure corresponding to twins.

A study of the microelastic free energy plot of Figure 12 yields interesting conclusions regarding the possibility that a rise in temperature will induce a f.c.c. → b.c.c. phase transformation in those metals in which the b.c.c. structure is almost, but not quite, mechanically unstable. In such metals the free energy, when plotted as in Figure 12, will have a shallow minimum in the b.c.c. position at 0° K., a minimum which lies, however, above that at the f.c.c. position. As the temperature is now raised, a shallow minimum will be associated with an unusually large amplitude of vibration of the (110) [$\bar{1}$10] shear strain coordinate and hence with an unusually large entropy. An unusually large entropy implies, however, a rapid decrease of free energy with rise of temperature. One therefore anticipates that, as the temperature rises, the b.c.c. minimum will be lowered more rapidly than will the f.c.c. minimum and may eventually become even lower than the f.c.c. minimum. Such considerations have led to a search for a transition of the b.c.c. metals to a f.c.c. structure at low temperatures. In this search Barrett[1] has discovered a f.c.c. modification of lithium at liquid-air temperatures.

Iron suffers a double transition—b.c.c. → f.c.c. at 910° C. and back again at 1400° C. The upper transformation may be interpreted in terms of the foregoing discussion, namely, in terms of the low value of the $(c_{11} - c_{12})/2$ coefficient likely to be associated with a b.c.c. structure. The lower transformation finds a ready interpretation in the fact that iron in the b.c.c. structure, but not in the f.c.c. structure, becomes spontaneously magnetized and that such magnetization is associated at low temperatures with a lowering of the free energy. According to this viewpoint, one would expect the lower transformation to occur at the Curie temperature. Actually, this transformation does not occur until about 150° C. above the Curie temperature. It is possible that this delay of 150° C. in the transformation arises from the lowering in the free energy associated with the local correlation between neighboring spins.

1. C. S. Barrett, *Phys. Rev.*, LXXII (1947), 245.

PART TWO

ANELASTICITY OF METALS

V

FORMAL THEORY OF ANELASTICITY

A. GENERALIZATIONS OF ELASTIC EQUATIONS

EARLY attempts were made, originally by O. Meyer[1] and W. Voigt,[2] to generalize the equations of the classical elasticity theory so as to include anelastic phenomena. The generalizations consisted essentially in regarding the stress components as a linear function of both strain and strain rates. Thus, for the simple case of tension, the tensile stress, σ, and the tensile strain, ϵ, were assumed to be related by the equation

$$\sigma = \alpha\epsilon + \beta\dot{\epsilon} \; . \tag{42}$$

Solids which supposedly obey equation (42) are known as "Voigt solids." As illustrated in Figure 16, such solids manifest an elastic aftereffect. It is evident that such solids will also manifest internal friction.

When a stress is suddenly applied to a Voigt solid, there is no instantaneous strain; but, as may be seen from equation (42), the strain gradually approaches an asymptotic value. Conversely, when the stress is suddenly removed, the solid suffers no instantaneous recovery, but the strain gradually disappears. This behavior is illustrated in Figure 16. While cork behaves in just such a manner,[3] all metals manifest an instantaneous strain upon sudden application of a stress and an instantaneous recovery upon removal of the stress.

When a Voigt solid is subject to oscillating stresses, the rate of dissipation of energy is proportional to the square of the frequency of oscillation, as may be seen from equation (42). Lord Kelvin[4] has observed that the rate of dissipation of energy increases less rapidly than the square of the frequency. It is therefore evident that the simple generalization of the elastic equations represented by equation (42) does not correspond to reality.

A clue as to a more successful generalization of the elastic equations may be obtained by representing a solid by an appropriate mechanical

1. "Zur Theorie der innere Reibung," *Jour. rein. u. angew. Math.*, LXXVIII (1874), 130.

2. "Über innere Reibung fester Körper, insbesondere der Metalle," *Ann. d. Phys.*, XLVII (1892), 671.

3. E. Schmid, *Gesundheits-Ing.*, XLVI (1923), 69.

4. Sir W. Thomson, *Math. and Phys. Papers*, III (Cambridge: Cambridge University Press, 1890), 27.

model, as was first done by Poynting and Thomson,[5] and then deriving the equation for this model. A mechanical model which represents the Voigt solid is shown in Figure 14. This model consists of a spring and dashpot in parallel, the latter having the property that its rate of displacement is proportional to the force acting upon it. A model which does not have the defects of the Voigt model may be obtained simply by placing a spring in series with the dashpot, as illustrated in Figure 15. Such a system will have an instantaneous displacement when a force is suddenly applied. The magnitude of this instantaneous displacement is determined

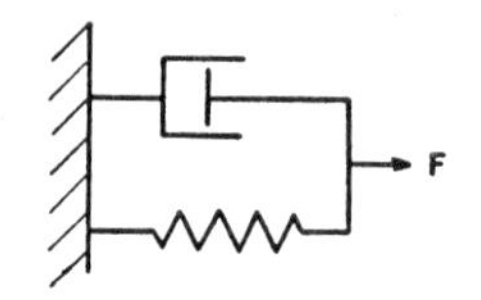

FIG. 14.—Mechanical model of a Voigt solid.

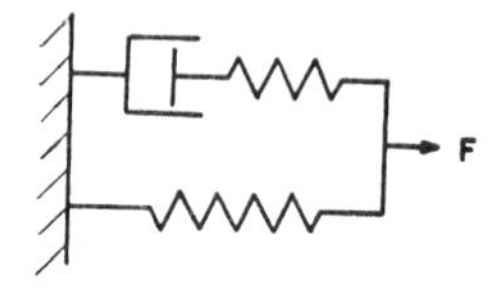

FIG. 15.—Mechanical model of a standard linear solid.

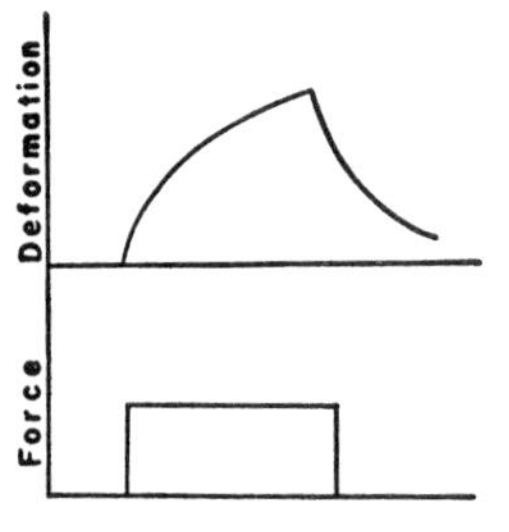

FIG. 16.—Mechanical behavior of a Voigt solid.

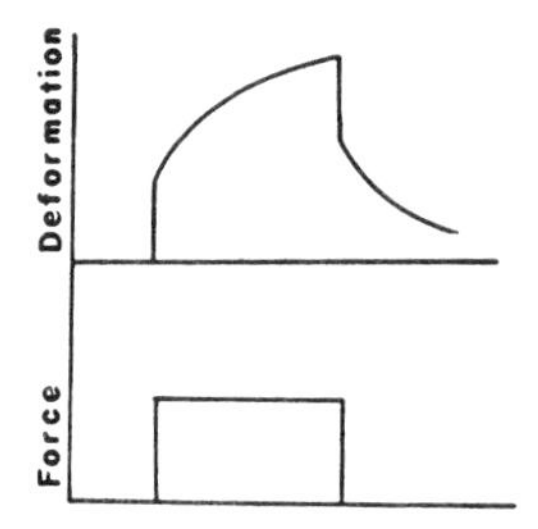

FIG. 17.—Mechanical behavior of a standard linear solid.

solely by the spring constants. As the force is maintained, the force across the dashpot is gradually relaxed by deformation therein, resulting in a gradual increase in the observed over-all deformation. Conversely, when the force is suddenly removed, the springs will suddenly release some of their stored energy, resulting in a partial instantaneous recovery. The complete release of all the energy in the springs must await the gradual relaxation of force across the dashpot. Such a system therefore manifests the general features of elastic after effects in real solids, illustrated in Figure 17. Further, the rate of energy dissipation by such a model during forced oscillation does not increase rapidly as the frequency increases. In fact, it may be seen that for large frequencies, when the displacement contributed by the dashpot is small compared with that contributed by

5. J. H. Poynting and J. J. Thomson, *Properties of Matter* (London: C. Griffin & Co., 1902).

its companion spring, the rate of energy dissipation is independent of frequency. The equation corresponding to the mechanical model of Figure 15 may be seen to have the form

$$a_1\sigma + a_2\dot{\sigma} = b_1\epsilon + b_2\dot{\epsilon} \,. \tag{43}$$

Since this is the most general linear homogeneous equation in stress, strain, and their first time derivatives, a solid which obeys equation (43) will be called a "standard linear solid."

While an elastic body subjected only to tensile stresses has only one constant, namely, the tensile elastic modulus, the corresponding standard linear solid has three essentially independent constants. Thus in the notation of equation (43) the three independent constants may be taken as a_2/a_1, b_1/a_1, b_2/a_1. In order to facilitate the understanding of the behavior of a standard linear solid, three new independent constants—τ_ϵ, τ_σ, M_R—will be introduced as follows:

$$\sigma + \tau_\epsilon\dot{\sigma} = M_R(\epsilon + \tau_\sigma\dot{\epsilon}) \,. \tag{44}$$

Here τ_ϵ is the time of relaxation of stress under conditions of constant strain. Thus suppose that both ϵ and $\dot{\epsilon}$ are zero. Then equation (44) reduces to

$$\sigma + \tau_\epsilon\dot{\sigma} = 0 \,,$$

which has the solution

$$\sigma(t) = \sigma(0)\, e^{-t/\tau_\epsilon} \,. \tag{45}$$

On the other hand, suppose a strain, ϵ_0, is suddenly applied at $t = 0$. The stress then relaxes with the relaxation time τ_ϵ to its equilibrium value $M_R\epsilon_0$. Thus the solution to equation (44) corresponding to this condition is

$$\sigma(t) = M_R\epsilon_0 + (\sigma_0 - M_R\epsilon_0)\, e^{-t/\tau_\epsilon} \,, \tag{46}$$

where σ_0 is the initial value of the stress. Since the final value of the ratio of stress to strain—the value after all relaxation has occurred—is M_R, this quantity is known as the "relaxed elastic modulus." The constant τ_σ is the time of relaxation of strain under conditions of constant stress. Thus, if both σ and $\dot{\sigma}$ are zero, ϵ obeys the equation

$$\epsilon + \tau_\sigma\dot{\epsilon} = 0 \,. \tag{47}$$

On the other hand, suppose a stress σ_0 is suddenly applied at $t = 0$. The strain then relaxes with the time of relaxation τ_σ to its equilibrium value, $M_R^{-1}\sigma_0$. Thus the solution to equation (44) corresponding to this condition is

$$\epsilon(t) = M_R^{-1}\sigma_0 + (\epsilon_0 - M_R^{-1}\sigma_0)\, e^{-t/\tau_\sigma} \,. \tag{48}$$

From the three independent constants, τ_ϵ, τ_σ, and M_R, certain new constants may be derived which have considerable physical significance. Suppose that in a very short time δt, the stress receives a finite increment $\Delta\sigma$. If we now integrate both sides of equation (44) with respect to time over the time interval δt, the first term in each side approaches zero as the time interval δt is made smaller and smaller. We are left with the following relation between the increments of stress, $\Delta\sigma$, and of strain, $\Delta\epsilon$:

$$\tau_\epsilon \Delta\sigma = M_R \tau_\sigma \Delta\epsilon .$$

The ratio $\Delta\sigma/\Delta\epsilon$ will be called the "unrelaxed elastic modulus" and will be denoted by M_U, since it gives the relation between changes in σ and in ϵ which occur so rapidly that no relaxation has time to take effect. Thus

$$\Delta\sigma = M_U \Delta\epsilon , \tag{49}$$

with M_U given by the equation

$$\frac{M_U}{M_R} = \frac{\tau_\sigma}{\tau_\epsilon} . \tag{50}$$

The deviation of the ratio M_R/M_U from unity may be used as a measure of the relative change in stress or in strain which may occur through relaxation. Examples of relaxation are given in Figures 18 and 19 for the cases in which M_R/M_U is 0.1 and 0.9, respectively. The first case is again illustrated in Figure 20, with time given on a logarithmic scale. In this figure it is particularly evident that the relaxation time of strain at constant stress is larger than is the relaxation time of stress at constant strain.

Solids are frequently investigated by dynamical methods. It is therefore of interest to find the relationship between stress and strain in the standard linear solid when these quantities are periodic. Toward this end we substitute in equation (44) the solutions

$$\sigma(t) = \sigma_0 e^{i\omega t}, \qquad \epsilon(t) = \epsilon_0 e^{i\omega t},$$

and endeavor to find the relation between σ_0 and ϵ_0. We obtain

$$(1 + i\omega\tau_\epsilon)\,\sigma_0 = M_R (1 + i\omega\tau_\sigma)\,\epsilon_0 \tag{51}$$

or

$$\sigma_0 = \mathfrak{M}\epsilon_0 ,$$

where the complex modulus $\mathfrak{M}$ is given by

$$\mathfrak{M} = \frac{1 + i\omega\tau_\sigma}{1 + i\omega\tau_\epsilon} M_R . \tag{52}$$

The angle δ by which strain lags behind stress is of particular interest, since its tangent is frequently used as a measure of internal friction. Upon observing that

$$\tan\,\delta = \frac{\text{Imaginary part of } \mathfrak{M}}{\text{Real part of } \mathfrak{M}} ,$$

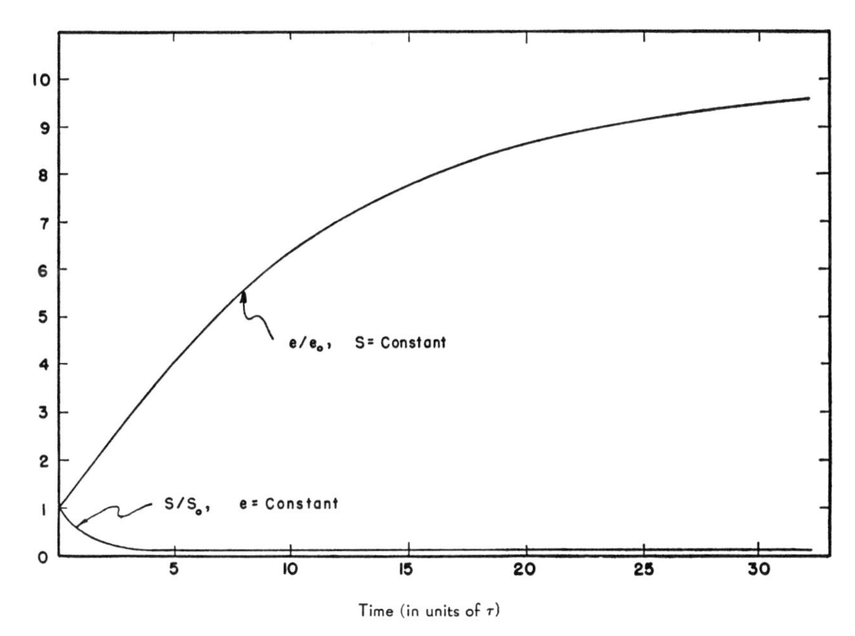

FIG. 18.—Relaxation of stress and strain, $M_R / M_U = 0.1$. Linear time scale

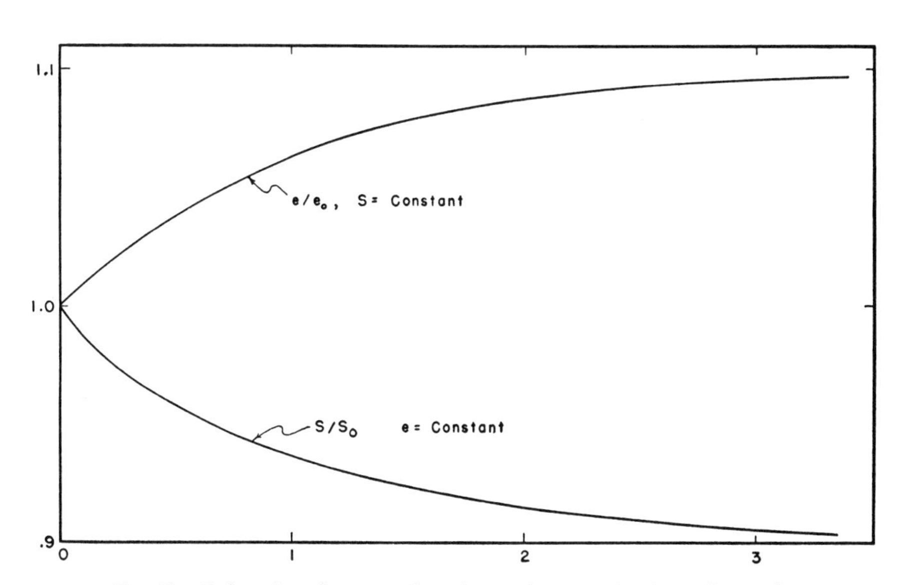

FIG. 19.—Relaxation of stress and strain, $M_R/M_U = 0.9$. Linear time scale

we obtain

$$\tan\,\delta = \frac{\omega\,(\tau_\sigma - \tau_\epsilon)}{1 + \omega^2\,(\tau_\sigma \tau_\epsilon)}\,. \tag{53}$$

In order to represent the variation of internal friction with angular frequency in as vivid a manner as possible, we shall introduce the geometric mean of the two times of relaxation, τ_ϵ and τ_σ,

$$\bar{\tau} = (\tau_\epsilon \tau_\sigma)^{1/2}, \tag{54}$$

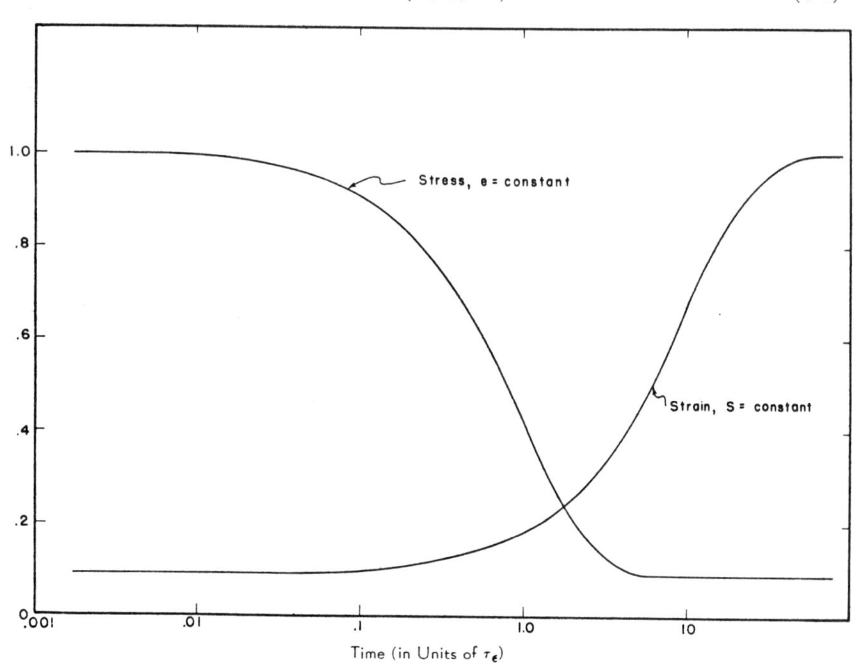

FIG. 20.—Relaxation of stress and strain for special case of $M_R/M_U = 0.1$. Logarithmic time scale.

and the geometric mean of the two moduli, M_U and M_R,

$$\bar{M} = (M_U M_R)^{1/2}. \tag{55}$$

Making use of equation (50), we then transform equation (53) to

$$\tan\,\delta = \frac{M_U - M_R}{\bar{M}} \cdot \frac{\omega\bar{\tau}}{1 + (\omega\bar{\tau})^2}\,. \tag{56}$$

The first factor in equation (56) is essentially the relative difference in the relaxed and the unrelaxed elastic moduli. The second factor gives the frequency variation of tan δ. This factor has a maximum when the product, $\omega\bar{\tau}$, is unity. When this factor is plotted against the logarithm of this product, it is seen to be a symmetrical function of $\omega\bar{\tau}$, with the general characteristics of an error curve. The maximum value of this second factor is $\frac{1}{2}$, and hence

$$(\tan \delta)_{\max} = \frac{M_U - M_R}{2\overline{M}}.$$

It is also of interest to have some real measure of the ratio of stress to strain, a ratio which will be given the symbol M_ω. The definition of M_ω is not unique. Thus we could define it as the magnitude of $\mathfrak{M}$, as the real part of $\mathfrak{M}$, or as the reciprocal of the real part of $\mathfrak{M}^{-1}$. The latter definition will be adopted simply because of the symmetrical expression to which it leads. According to this definition, M_ω is the ratio of the stress to that part of the strain which is in phase with the stress. From equation (52) we find that this definition leads to

$$M_\omega = \frac{1 + \omega^2 \tau_\sigma^2}{1 + \omega^2 \tau_\sigma \tau_\epsilon} M_R.$$

Upon using equation (50), we find M_ω may be written as

$$M_\omega = M_U - \frac{M_U - M_R}{1 + (\omega\overline{\tau})^2}.$$

In the limits of low and high frequency M_ω reduces to the following limits:

$$M_\omega = \begin{cases} M_R, & \omega\overline{\tau} \ll 1, \\ M_U, & \omega\overline{\tau} \gg 1. \end{cases}$$

The precise manner in which M_ω varies between these two limits is shown in Figure 21. The variation of M_ω is seen to be a maximum when the product $\omega\overline{\tau}$ is unity, just where $\tan \delta$ is also a maximum. It may likewise be seen from this figure that M_ω has effectively reached its two extreme limits at frequencies where $\tan \delta$ is still appreciable. This effective narrowness of the transition range of M_ω, as compared to the range in which $\tan \delta$ is appreciable, is a manifestation of the difference between the asymptotic expansions of these two quantities, as illustrated in the following equations:

$$\left.\begin{aligned} M_U - M_\omega &= \frac{M_U - M_R}{(\omega\overline{\tau})^2} \\ \tan \delta &= \frac{M_U - M_R}{\overline{M}\,\omega\overline{\tau}} \end{aligned}\right\} \omega\overline{\tau} \gg 1,$$

$$\left.\begin{aligned} M_\omega - M_R &= (M_U - M_R) \cdot (\omega\overline{\tau})^2 \\ \tan \delta &= \frac{M_U - M_R}{\overline{M}} \cdot (\omega\overline{\tau}) \end{aligned}\right\} \omega\overline{\tau} \ll 1.$$

4 /

B. BOLTZMANN'S SUPERPOSITION PRINCIPLE

While the standard linear solid discussed above has certain general features in common with actual solids, it does not reproduce precisely the behavior of any real metal. In an attempt to formulate a theory which corresponds more closely to reality, two paths may be followed. On the one hand, one may attempt to describe the mechanical behavior by means of a differential equation, similar to equation (47) for the standard

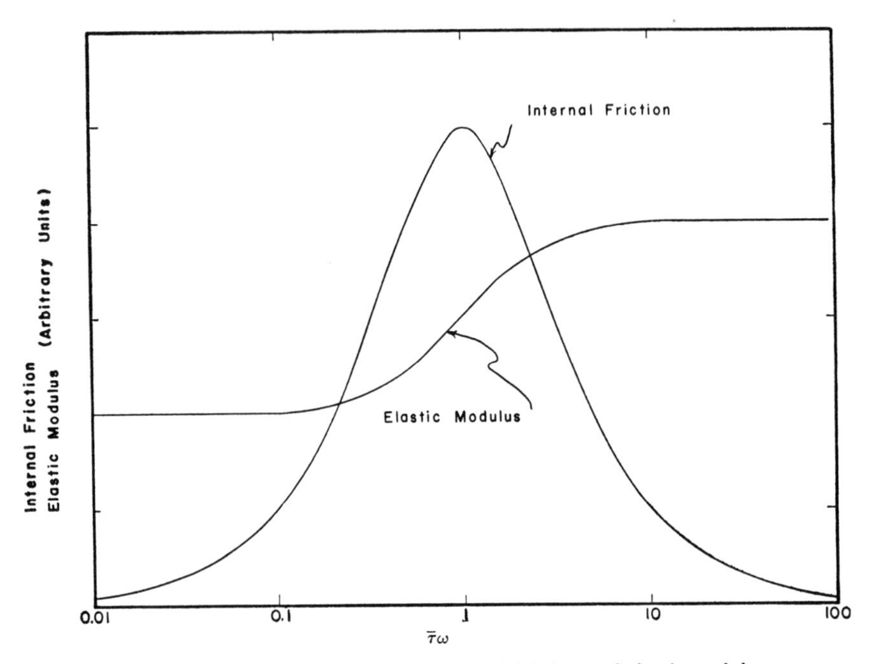

FIG. 21.—Frequency dependence of internal friction and elastic modulus

linear solid but containing higher derivatives of stress and of strain. This procedure has recently been discussed by Alfrey.[6] On the other hand, one may abandon the attempt to derive a differential equation relating stress and strain and develop all the conclusions derivable from the assumption that the true relation between stress and strain is linear. This latter method was introduced by Boltzmann[7] and will be discussed in detail in this section.

If the fundamental relations between stress and strain are linear in

6. T. Alfrey, "Non-homogeneous Stresses in Visco-elastic Media," *Quart. Jour. Appl. Math.*, II (1944), 113; and "Methods of Representing the Properties of Visco-elastic Materials," *ibid.*, III (1945–46), 143.

7. L. Boltzmann, "Zur Theorie der elastische Nachwirkung," *Ann. d. Phys.*, VII (1876), 624; also *Sitzungsb. Akad. Wiss. Wien. Ber.*, LXX (1874), 275.

stress and strain and in their higher time derivatives, then the solutions satisfy the principle of superposition. Thus, let $D_1(t)$ be the deformation produced by the force $F_1(t)$ acting alone and $D_2(t)$ be the deformation produced by the force $F_2(t)$ acting alone, then this principle states that $D_1(t) + D_2(t)$ is the total deformation produced by the forces $F_1(t)$ and $F_2(t)$ acting together. Expressed in more formal language, the superposition principle states:

If $D_1(t)$ is produced by $F_1(t)$
and $D_2(t)$ is produced by $F_2(t)$,
then $D_1(t) + D_2(t)$ is produced by $F_1(t) + F_2(t)$.

Conversely, the deformation may be regarded as the independent variable, the applied force as the dependent variable. For example, the deformation may be varied according to a predetermined schedule, and the force measured which is necessary to produce this deformation. In this case the superposition principle states:

If $F_1(t)$ is required for $D_1(t)$
and $F_2(t)$ is required for $D_2(t)$,
then $F_1(t) + F_2(t)$ is required for $D_1(t) + D_2(t)$.

The superposition principle may be formulated in a still more concise manner. Toward this end we define the quantity $\delta(t)$, which will be called the "creep function," as the deformation resulting from the sudden application at $t = 0$ of a constant force of magnitude unity. Now, if during the time interval from t to $t + dt$ the force changes from F to $F + \dot{F}dt$, we may consider that during this time interval a constant force has been applied of magnitude $\dot{F}dt$. According to this viewpoint, the deformation at any instant t is the result of a continuous series of constant forces previously applied. Thus, according to this viewpoint, the superposition principle states:

$$D(t) = \int_{-\infty}^{t} \delta(t-t') \dot{F}(t') dt' . \tag{57}$$

Conversely, as before, we may regard the deformation as a specified function of time. Toward this end we define the quantity $f(t)$, which will be called the "stress function," as the force which must be applied in order that the deformation may suddenly change at $t = 0$ from zero to unity and remain unity thereafter. By applying the same reasoning as that used for equation (57), we derive

$$F(t) = \int_{-\infty}^{t} f(t-t') \dot{D}(t') dt' . \tag{58}$$

The anelastic behavior of a solid may be regarded as completely specified either by $\delta(t)$ or by $f(t)$. In order to obtain a relation between these two quantities we shall apply equation (57) to the case in which

$$D(t) = \begin{cases} 0\,, & t<0\,, \\ 1\,, & t>0\,, \end{cases}$$

and hence where also

$$F(t) = \begin{cases} 0\,, & t<0\,, \\ f(t)\,, & t>0\,. \end{cases}$$

After $t = 0$, the left-hand side of equation (57) is equal to unity, while the right-hand side may most conveniently be separated into two terms. The first term is the integral over the time interval in the immediate vicinity of $t = 0$ during which $D(t)$ is changing from 0 to 1. The second term is the integral after $D(t)$ has attained the value unity. We thus find

$$1 = \delta(t)\, f(0) + \int_0^t \delta(t-t')\, \dot{f}(t')\, dt'\,. \tag{59}$$

This equation relates the two quantities $\delta(t)$ and $f(t)$. The precise evaluation of one function in terms of the other would be very difficult, requiring iterated numerical integrations. Certain useful relations may, nevertheless, be derived without such numerical integrations. The two factors of the integrand in equation (59) are plotted in Figure 22 as a function of t' for typical examples. We see that $\delta(t - t')$ differs appreciably from $\delta(t)$ only when $\dot{f}$ is very small. It will therefore be advantageous to write

$$\delta(t-t') \qquad \text{as} \qquad \delta(t) - \{\delta(t) - \delta(t-t')\}\,.$$

The first term then contributes the major part of the integral, and, furthermore, its integration may be carried out directly, leading to the following relation between the creep and stress functions:

$$\delta(t)\, f(t) = 1 + \int_0^t \{\delta(t) - \delta(t-t')\}\, \dot{f}(t')\, dt'\,. \tag{60}$$

Now the first factor in the integrand is always positive, and the second factor is always negative, so that the creep and stress functions satisfy the following inequality:

$$\delta(t)\, f(t) \leqslant 1\,. \tag{61}$$

It may be seen from equation (60) that the equality sign is valid at very short times. If $\delta(t)$ approaches a limiting value, the equality sign is also valid at very long times, the inequality being appropriate only for intermediate times.

As an example of the above discussion, we shall evaluate $\delta(t)$, $f(t)$, and the product $\delta(t)f(t)$ for the standard linear solid. From equations (49) and (50) we obtain

$$\delta(t) = M_R^{-1} - (M_R^{-1} - M_U^{-1})\, e^{-(M_R/M_U)\, t/\tau}, \tag{62}$$

$$f(t) = M_R + (M_U - M_R)\, e^{-t/\tau}. \tag{63}$$

These functions and their product are plotted in Figure 23 for the special case of $M_U/M_R = 10$. The product $\delta(t)f(t)$ for this particular case reaches a minimum value of 0.65.

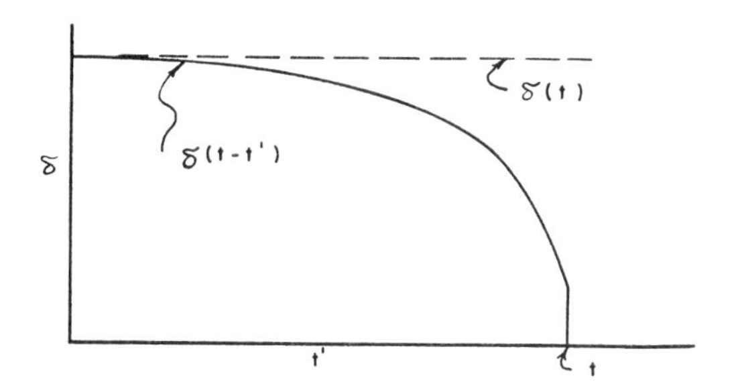

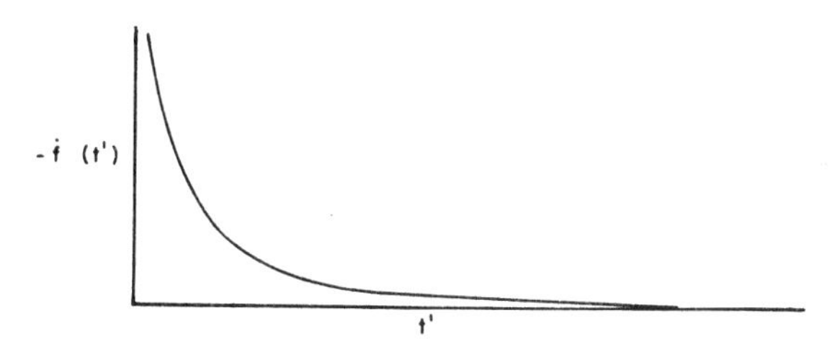

FIG. 22.—Interpretation of integrand of equation (19)

Anelastic experiments are frequently made in which a periodic displacement or force is applied. We shall now examine how the results of such experiments are related to the creep and relaxation experiments.

In order to analyze the force required for a periodic deformation, we shall find it convenient to re-write equation (58) in another form. We first integrate by parts, obtaining

$$F(t) = f(0)\, D(t) + \int_{-\infty}^{t} \dot{f}(t - t')\, D(t')\, dt', \tag{64}$$

and then substitute τ for $t - t'$, obtaining

$$F(t) = f(0)\, D(t) + \int_{0}^{\infty} \dot{f}(\tau)\, D(t - \tau)\, d\tau. \tag{65}$$

From this equation we find that the force required for the periodic deformation,

$$D(t) = D_0 \sin \omega t ,$$

is given by

$$F(t) = D_0 \left\{ f(0) + \int_0^\infty \dot{f}(t) \cos \omega t \, dt \right\} \sin \omega t - D_0 \left\{ \int_0^\infty \dot{f}(t) \sin \omega t \, dt \right\} \cos \omega t . \quad (66)$$

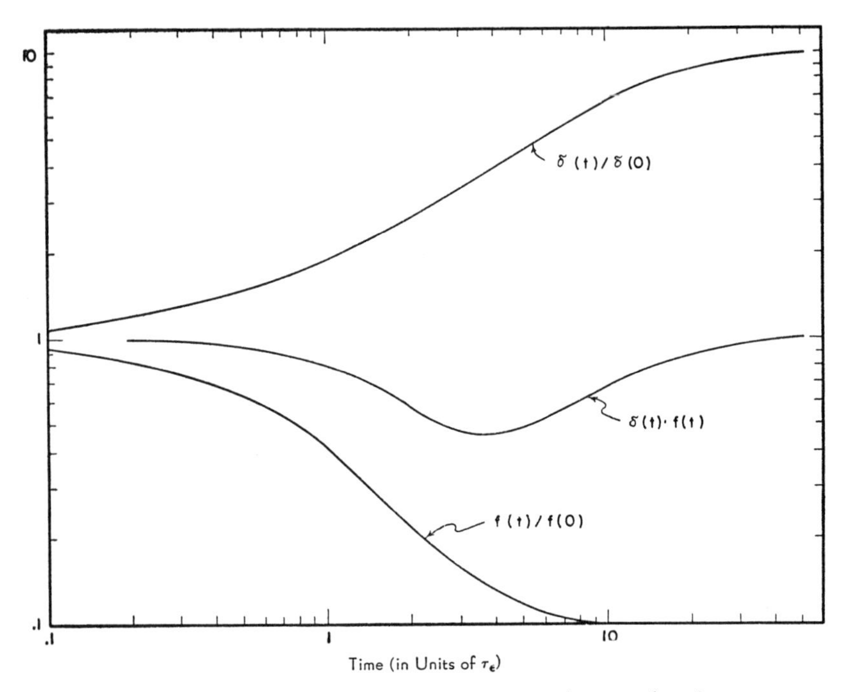

FIG. 23.—Example of relation between creep and stress relaxation

From equation (66) we see that a periodic displacement requires a periodic force. Conversely, a periodic force gives rise to a periodic displacement. The corresponding stress-strain diagram is therefore an ellipse, and the relation between force and deformation may be represented by a complex modulus $\mathfrak{M}$. The real part of this complex modulus will be called the "dynamic modulus" and will be denoted by $M(\omega)$.

From equation (66) we may derive, in certain important cases, the relation between the dynamic modulus and the stress function. The important case for which this relation may be derived is that in which $d \ln f / d \ln t$ is small compared with unity and varies only slightly with $\ln t$. An example of a stress function satisfying this condition is given in Figure 24. In this figure $d \ln f / d \ln t$ has the constant value 0.20, and f decreases from 0.8 to 0.5 as t increases from 0.1 to 1,000. The salient features of such a plot of $f(t)$ on a linear time scale are independent of the time units. Thus, irrespective of whether the time scale extends from 0 to 10, from 0 to 100, or from 0 to 1,000 seconds, f drops 30 or 40 per

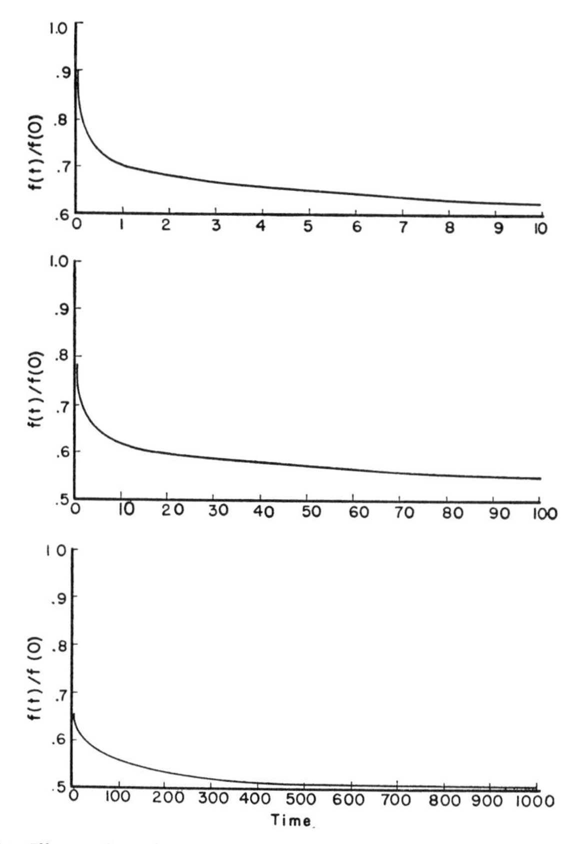

FIG. 24.—Illustration of a typical stress relaxation plotted on a linear scale

cent during the first tenth of the time interval and 8 per cent, at most, during the remaining nine-tenths of the interval. From equation (66) the dynamic modulus is given by

$$M(\omega) = f(0) + \int_0^\infty \dot{f}(t) \cos \omega t \, dt .$$

When integrated by parts, this equation becomes

$$M(\omega) = \int_0^\infty f\left(\frac{x}{\omega}\right) \sin x \, dx .$$

The contributions to the integral from the ranges $0-\pi$, $\pi-2\pi$, $2\pi-3\pi$, . . . , are of alternate signs and of slowly decreasing magnitude. A very good approximation to the sum of this alternating series is given by the integral extended over the first half of the first range. Thus

$$M(\omega) \simeq \int_0^{\pi/2} f\left(\frac{x}{\omega}\right) \sin x \, dx .$$

As may be seen from Figure 24, a good approximation to this integral is obtained by regarding $f(x/\omega)$ as a constant, with its value in the center of the range over which the integration extends. We are therefore led to the following relation between the dynamic modulus and the stress relaxation function:

$$M(\omega) \simeq f\left(\frac{P}{8}\right), \tag{67}$$

where P is the period of oscillation, $2\pi/\omega$.

From equation (66) we may also derive, for the above-discussed case, the relation between the stress function and the internal friction. As has previously been mentioned, the internal friction is the tangent of the angle δ by which the deformation lags behind the applied force in periodic oscillation, and it is, when small, proportional to the relative energy dissipated per cycle. The tangent of the angle by which the deformation lags behind the force is the ratio of the coefficient of $\cos \omega t$ to the coefficient of $\sin \omega t$ in equation (66), and hence

$$\tan \delta \simeq \frac{\int_0^\infty \dot{f}(t) \sin \omega t \, dt}{f\left(\frac{P}{8}\right)} . \tag{68}$$

If we now write the integrand of the numerator as $(df/d \ln t)$ $(\sin \omega t)/t$, the first factor may be regarded as essentially constant. Integration over the remainder of the integrand gives the numerical factor $\pi/2$. Now the major part of this integral, 87 per cent, comes from the integration over the range $0 < \omega t < \pi/2$. The essentially constant factor $df/d \ln t$ can therefore reasonably be assigned its value in the middle of this range. We thus obtain the following relation between the internal friction and the stress function:

$$\tan \delta(\omega) \simeq -\frac{\pi}{2}\left(\frac{d \ln f}{d \ln t}\right)_{t=P/8} . \tag{69}$$

Equations (60) or (61), (67), and (69) constitute three independent relations between the following four functions: creep, stress, dynamic modulus, and internal friction. The equality sign in equation (61) and equations (67) and (69) are good approximations for the cases typified by Figure 24. Three more approximate equations could therefore be derived from the above, one between $M(\omega)$ and $\delta(t)$ by substitution of δ^{-1} for f in equation (67); one between $\tan \delta(\omega)$ and δ by substituting δ^{-1} for f in equation (68); and, finally,

$$\tan \delta(\omega) \simeq -\frac{\pi}{2} \frac{d \ln M(\omega)}{d \ln \omega}, \qquad (70)$$

obtained by combining equations (67) and (69). A beautiful experimental demonstration of the validity of equations (67)–(70) has been presented by Kê[8] in his study upon the viscosity of grain boundaries.

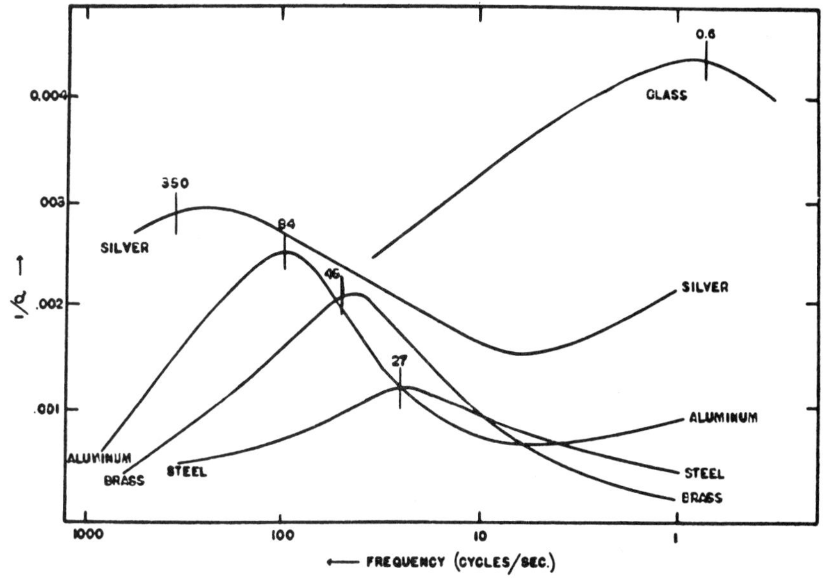

FIG. 25.—Internal friction measured by Bennewitz and Rötger. The calculated values of the maxima are marked by vertical lines.

C. RELAXATION SPECTRUM

A particularly simple example of anelasticity was discussed in Section A. In this example a certain fraction of the initial stress relaxes in an exponential manner when the strain is maintained constant. Conversely, when the stress is maintained constant, the strain approaches an asymptotic value in an exponential manner. In chapter vii we shall see that many examples occur in nature in which the creep and stress relaxations obey these exponential laws. One is therefore tempted to regard all examples of anelastic phenomena as the superposition of elementary processes in which the stress relaxes exponentially. This point of view was adopted many

8. T. S. Kê, *Phys. Rev.*, LXXI (1947), 533.

years ago by J. J. Thomson[9] and by E. Wiechert.[10] In previous discussions of this point of view it has been assumed that the stress which relaxes is only a small percentage of the original stress. The recent observations of W. A. West[11] show that the major fraction of the stress may, in fact, relax. It is therefore necessary to re-examine this viewpoint.

In discussing various types of stress relaxation, it is desirable to have some dimensionless measure of the total stress relaxation. Expressed in terms of the unrelaxed and relaxed moduli, one could use $(M_U - M_R)/M_U$ or $(M_U - M_R)/M_R$ or $(M_U - M_R)/\sqrt{M_R M_U}$. In Section A of chapter vii we shall see that, in the case of simple exponential relaxations, the second of the above measures of relaxation is most appropriate, since it may readily be expressed in terms of the physical constants of the system. We shall therefore define the relaxation strength of a simple exponential relaxation by the quantity

$$\Delta = \frac{M_U - M_R}{M_R}. \tag{71}$$

In most experiments the pertinent modulus is either Young's modulus E, the rigidity modulus G, or the bulk modulus K. The measured relaxation strength will, in general, depend upon which modulus governs the relation between stress and strain. A subscript will therefore be given Δ to denote the type of experiment to which the relaxation strength refers. Thus

$$\Delta_E = \frac{E_U - E_R}{E_R}, \tag{72}$$

$$\Delta_G = \frac{G_U - G_R}{G_R}, \tag{73}$$

$$\Delta_K = \frac{K_U - K_R}{K_R}. \tag{74}$$

Generally no relaxation occurs under a pure hydrostatic pressure; thus in these cases Δ_K is exactly zero. When Δ_K is zero, a relation exists between Δ_E and Δ_G. From the relation[12] between E, G, and K, namely,

$$E = \frac{9KG}{3K + G},$$

one derives

$$\Delta_G = 3\,\frac{G_U}{E_U}\,\Delta_E.$$

9. "On Residual Effects," in his *Applications of Dynamics to Physics and Chemistry* (London: Macmillan & Co., 1888), chap. viii.

10. "Über elastische Nachwirkung" (inaugural dissertation [Königsberg i. Pri., 1889]); also "Gesetze der elastische Nachwirkung für konstant Temperature," *Ann. d. Phys.*, L (1893), 335.

11. "Elastic After-effects in Iron Wires from 20° to 550° C.," *Trans. A.I.M.E.*, CLXVII (1946), 192.

12. R. V. Southwell, *Theory of Elasticity* (Oxford: Clarendon Press, 1936), p. 307.

Upon using the following formula relating G_U and E_U in terms of Poisson's ratio, σ:

$$E_U = 2\,(1+\sigma)\,G_U\,,$$

we obtain

$$\Delta_G = \frac{3}{2\,(1+\sigma)}\,\Delta_E\,. \tag{75}$$

For the usual range of σ, Δ_G is about 15 per cent larger than Δ_E.

The relaxation strength of a single crystal will depend upon its crystallographic orientation, as well as upon the type of deformation to which it is subjected. If we assume that a cubic crystal undergoes no relaxation under a pure hydrostatic pressure, it will have two relaxation strengths, corresponding to the relaxation of the two shear coefficients, c_{44} and $(c_{11}-c_{12})/2$. These two relaxation strengths will be denoted by δ and δ', respectively. In order to compute Δ_E and Δ_G in terms of these two relaxation strengths, we need the following formulae for the tensile and torsion moduli:[13]

$$E^{-1} = s_{11} - [2\,(s_{11}-s_{12}) - s_{44}]\,\phi\,, \tag{75a}$$

$$G^{-1} = s_{44} + 2\,[2\,(s_{11}-s_{12}) - s_{44}]\,\phi\,, \tag{75b}$$

where

$$\phi = \gamma_1^2\gamma_2^2 + \gamma_2^2\gamma_3^2 + \gamma_3^2\gamma_1^2\,,$$

γ_1, γ_2, γ_3 being the direction cosines of the specimen's axis with respect to the principal axes of the crystal. We shall also need the following formulae relating the elastic coefficients (c's) to the elastic constants (s's):

$$s_{44} = \frac{1}{c_{44}}\,, \tag{75c}$$

$$s_{11} - s_{12} = \frac{1}{c_{11}-c_{12}}\,, \tag{75d}$$

$$s_{11} + 2\,s_{12} = \frac{1}{c_{11}+2\,c_{12}}\,, \tag{75e}$$

equations which are identical to equation (22) of Part I. From equations (75*a*)–(75*e*) we find directly that, for cubic crystals,

$$\Delta_E = \frac{\dfrac{C\,\delta'}{3} - (C\,\delta' - C'\,\delta)\,\phi}{\dfrac{c_{44}\,(c_{11}+c_{12})}{2\,(c_{11}+2\,c_{12})} - (C-C')\,\phi}\,, \tag{75f}$$

$$\Delta_G = \frac{C'\,\delta + 2\,(C\,\delta' - C'\,\delta)\,\phi}{C' + 2\,(C-C')\,\phi}\,, \tag{75g}$$

13. Cf. eq. (19) of Part I; also E. Schmid and W. Boas, *Kristallplastizität* (Berlin: Springer, 1935), p. 23.

where C and C' denote the two shear coefficients, namely,

$$C = c_{44} ,$$

$$C' = \frac{c_{11} - c_{12}}{2} .$$

In deriving equations (75*f*) and (75*g*), we have made use of the equation

$$\delta \equiv \frac{(c_{44})_U - (c_{44})_R}{(c_{44})_R} = \frac{(s_{44})_R - (s_{44})_U}{(s_{44})_U}$$

and of a similar equation for δ'.

If the total relative stress relaxation of a specimen is much less than unity, the stress relaxations due to different sources may be regarded as independent of one another. A relaxation spectrum may then be defined unambiguously as the plot of the distribution function, $\Delta(\tau)$, against $ln\ \tau$, where $\Delta(\tau) \cdot d\ ln\ \tau$ is the contribution to the total relaxation strength of those simple exponential relaxations whose times of relaxation lie within the range $d\ ln\ \tau$ at τ. The distribution function, $\Delta(\tau)$, satisfies, by definition, the equation

$$\delta = \int_{-\infty}^{\infty} \Delta(\tau)\, d\ ln\ \tau . \tag{76}$$

The relative stress which yet remains to be relaxed by the simple exponential relaxations in the range $d\ ln\ \tau$ at τ is

$$\Delta(\tau)\, e^{-t/\tau} d\ ln\ \tau .$$

The distribution function, $\Delta(\tau)$, may therefore, at least in principle, be determined from stress-relaxation experiments through the integral equation

$$\frac{\sigma(t) - \sigma(\infty)}{\sigma(\infty)} = \int_{-\infty}^{\infty} \Delta(\tau)\, e^{-t/\tau} d\ ln\ \tau . \tag{77}$$

Again, the contribution to internal friction arising from the simple exponential relaxations with relaxation times in the range $d\ ln\ \tau$ at τ is, from equation (53),

$$\Delta(\tau) \frac{\omega\tau}{1 + \omega^2\tau^2}\, d\ ln\ \tau .$$

The distribution function may therefore, in principle, be determined from internal-friction measurements through the equation

$$\tan\ \delta(\omega) = \int_{-\infty}^{\infty} \Delta(\tau) \frac{\omega\tau}{1 + \omega^2\tau^2}\, d\ ln\ \tau . \tag{78}$$

As has been mentioned above, the distribution function, $\Delta(\tau)$, may be computed in principle *either* from a knowledge of the stress-relaxation function, $\sigma(t)/\sigma(0)$, by means of equation (77) or from a knowledge of the internal friction, tan δ, as a function of frequency by means of equation (78). In each case the determination of $\Delta(\tau)$ involves the inversion of an integral. This inversion is practicable only when the stress relaxation and the internal friction are known analytic functions of time or frequency, respectively. In such cases standard inversion methods may be adopted, as has been recently discussed by Gross.[14] In such inversions small errors in the stress relaxation or internal friction functions would be reflected as relatively large errors in the distribution function, $\Delta(\tau)$.

Because of this extreme sensitivity of $\Delta(\tau)$ to small errors in observation, it has not seemed to the author profitable to express experimental results in terms of this distribution function. Rather, it has appeared wiser to express experimental results directly as tan δ. A plot of tan δ against the logarithm of time may then be regarded as the relaxation spectrum, as is illustrated by the Frontispiece.

14. B. Gross, "On Creep and Relaxation," *Jour. Appl. Phys.*, XVIII (1947), 212.

VI

MEASUREMENT OF RELAXATION SPECTRUM

A. RELATION BETWEEN MEASURES OF INTERNAL FRICTION

IN THE determination of the relaxation spectrum, namely, in the determination of the internal friction, tan δ, as a function of frequency, the angle of phase lag δ could be measured directly. This method must be used if the internal friction is comparable to, or greater than, unity. In the past only experiments have been reported in which the internal friction is small compared to unity. In such cases a variety of indirect methods may be employed to determine the internal friction.

One indirect method of measuring the internal friction is to observe the amplitude of vibration under conditions of forced oscillation, the frequency of the impressed force being slowly varied while its amplitude remains constant. The amplitude of vibration will be a maximum when the impressed frequency is equal to a critical resonance frequency, ν, of the specimen and decreases essentially to zero as the impressed frequency either increases or decreases away from this critical frequency. Let $\Delta\nu$ be the change in impressed frequency necessary to change the amplitude from half-maximum on one side of the maximum to half-maximum on the other side. It will then be shown below that, provided that tan δ is a small-order quantity, at the resonance frequency

$$\tan \delta = \frac{\Delta\nu}{\sqrt{3}\nu}. \tag{79}$$

A second indirect method of measuring internal friction involves the measurement of the energy, ΔE, dissipated per cycle of vibration. When tan δ is a small-order quantity, one may define the energy of vibration, E, as the strain energy where the strain is at a maximum. When the strain is not at a maximum, part of the energy E is in the form of kinetic energy. The ratio $\Delta E/E$ is commonly called the "specific damping capacity." It is shown below that, in the case of small tan δ,

$$\tan \delta = \frac{1}{2\pi} \cdot \frac{\Delta E}{E}. \tag{80}$$

A third indirect method of measuring internal friction is to observe the gradual decay in amplitude of vibration during free oscillation. The logarithmic decrement (log. dec.) is then defined as the natural logarithm of the ratio of successive amplitudes. It is shown below that, in the case of small $\tan\delta$,

$$\tan\delta = \frac{\text{log. dec.}}{\pi}. \tag{81}$$

It is evident that the above relation must become invalid for large values of the internal friction, for then the internal damping of the specimen itself will overdamp the vibration, resulting in aperiodic motion. The author[1] has shown that critical damping occurs when δ is 53°, provided that the specimen has only a single pair of relaxation times, τ_ϵ and τ_σ. In practice it is often convenient to observe the time, τ_n, required for the amplitude of vibration to decay to one-nth of its original value. If ν is the frequency of vibration, it is then evident from equation (81) that

$$\tan\delta = \frac{\ln n}{\pi\nu\tau_n}. \tag{82}$$

We shall now derive the theoretical relations given in equations (79)–(81). In order to derive the first of these equations, we must write the equation of motion of the specimen as a whole. This is particularly simple when an auxiliary inertia member is used of such magnitude that the inertia of the specimen itself may be neglected. Let I be the inertia of the auxiliary member, D the deformation of the specimen, f the "force" with which the auxiliary member acts upon the specimen, and F the external "force" which acts upon the auxiliary member. Thus in the case of torsion, I is the moment of inertia of the auxiliary member, D is the angle of twist, and f and F are torques. The equation of motion of the auxiliary member is then

$$I\ddot{D} = -f + F. \tag{83}$$

In the case of periodic motion we may consider D, f, and F to vary with time as

$$D = D_0 e^{i\omega t}, \text{ etc.} \tag{84}$$

Deformation and force may then be related by

$$\mathrm{F} = \mathfrak{M} D, \tag{85}$$

where the complex modulus $\mathfrak{M}$ may be written as

$$\mathfrak{M} = M(1 + i\tan\delta). \tag{86}$$

Upon substituting equations (84)–(86) in equation (83), we obtain

$$D = \frac{\frac{F}{M}}{\left(1 - \frac{I\omega^2}{M}\right) + i\tan\delta}.$$

1. C. Zener, *Jour. Appl. Phys.*, XVIII (1947), 1022.

The amplitude of the deformation is a maximum at the resonance angular frequency,

$$\omega_0 = \left(\frac{M}{I}\right)^{1/2}.$$

This amplitude is half its maximum value when

$$1 - \frac{\omega^2}{\omega_0^2} = \pm\sqrt{3}\tan\delta\,.$$

The difference of the two values of ω which satisfy this equation, $\Delta\omega$, is seen to be given by

$$\frac{\Delta\omega}{\omega_0} = \sqrt{3}\tan\delta\,,$$

in the case in which $\tan\delta$ is small. This equation is identical with equation (79). In cases in which an auxiliary inertia member is not employed, the proof of equation (79) follows in a similar manner, except that a different interpretation[2] must be assigned to the quantities D, f, F, and I. In particular, D is the coefficient of a normalized characteristic displacement vector in the expansion of the displacement vector in terms of these characteristic vectors. Analogous interpretations are assigned to f and F, while I becomes the density of the material.

We shall now derive equation (80). The rate at which energy is dissipated per unit volume is $\overline{\sigma\dot{\epsilon}}$. The energy dissipated per cycle throughout the specimen is therefore

$$\Delta E = \frac{2\pi}{\omega}\int \overline{\sigma\dot{\epsilon}}\, d\,v\,.$$

Upon setting

$$\sigma = \sigma_0\cos\omega t\,,$$
$$\epsilon = \epsilon_0\cos(\omega t - \delta)\,,$$

we obtain

$$\Delta E = \pi\sin\delta\int \epsilon_0\sigma_0 d\,v\,. \tag{87}$$

On the other hand, when $\tan\delta$ is small, we may take the energy of vibration per unit volume as $\frac{1}{2}\sigma_0\epsilon_0$. The energy of vibration of the whole specimen is therefore

$$E = \tfrac{1}{2}\int \sigma_0\epsilon_0\, d\,v\,. \tag{88}$$

Upon comparing equations (87) and (88), we obtain

$$\frac{\Delta E}{E} = 2\pi\sin\delta\,,$$

which is identical with equation (80) in the case in which $\tan\delta$ is small.

2. C. Zener, *Phys. Rev.*, LII (1937), 230.

Equation (81) may be derived most readily directly from equation (80). In free oscillation the logarithmic decrement for the energy of vibration is simply $\Delta E/E$. Since both ΔE and E are proportional to the square of the amplitude of vibration, provided that this amplitude is small, the logarithmic decrement for amplitude (log. dec.) is just one-half that for energy; whence

$$\text{log. dec.} = \frac{1}{2}\frac{\Delta E}{E}.$$

B. MEASUREMENT OF INTERNAL FRICTION

The earliest measurements[3] of internal friction were made by the method of free decay. The specimen formed the suspension element of a ballistic galvanometer, and the logarithmic decrement was obtained by observing the rate of decay of the amplitude of vibration. This early method has recently been revised and used with great success by Kê.[4] When the specimen is in the form of a wire, the period of oscillation may readily be made of the order of a second, and so the observations may be made visually. When specimens of the general size of a standard tensile specimen are used, the periods of vibration are correspondingly shorter, so that photographic or other nonvisual means of recording must be adopted. Such equipment was first used by Föppl[5] and has recently been improved by Norton.[6]

The free oscillation method may also be employed without the use of an auxiliary inertia member. In this case the specimen is excited in one of its normal modes of vibration, and the decay of amplitude is then observed. In such cases the frequency is considerably higher than when an auxiliary inertia member is employed, and hence direct visual observation may no longer be employed. For merely relative values of the internal friction, the time of decay may be determined remarkably accurately by the ear, provided, of course, that the frequency is in the audible range. This audio method has been used with considerable success by Waller.[7] Electrical methods are commonly employed accurately to observe the decay of oscillation. Thus by measuring the time τ_n required for the amplitude to be reduced to one-nth of its original value, the internal friction may be computed from equation (82). On the other hand, electrical methods have

3. W. Weber, *Poggendorf's Ann.*, XXXIV (1837), 247.

4. T. S. Kê, *Phys. Rev.*, LXXI (1947), 533; LXXII (1947), 41; *Jour. Appl. Phys.*, XIX (1948), 285.

5. O. Föppl, *Ber. d. Werkstoffausschusses d. Ver. deutsch. Eisenhuttenleute*, No. 36, 1923.

6. J. Norton, *Rev. Sci. Instruments*, X (1939), 77.

7. Mary D. Waller, *Proc. Phys. Soc.*, L (1938), 144; *Proc. Roy. Soc. London*, CLVI (1936), 383; *Jour. Sci. Instruments*, XII (1935), 300.

been described[8] in which one measures directly the number of oscillations required for the amplitude to be reduced by a predetermined factor.

Rather closely related to the method of free oscillation is the method in which a high-frequency pulse is introduced at one end of a bar, and the amplitude of the pulse is observed as it suffers successive reflections at the ends. This method has been employed by Huntington.[9] On the other hand, this method suffers from the danger that the loss of intensity of the wave packet may be due largely to scattering rather than to true absorption, as, for example, scattering by the random orientation of crystallites observed by Mason.[10]

When resonance frequencies are employed, the time of decay in free oscillation is frequently too short to measure conveniently. In such cases it is customary to measure the width at half-maximum of a resonance curve. The internal friction is then computed by equation (79). The resonance and decay methods may be regarded as complementary to each other. The former method is more applicable to those cases in which the width at half-maximum is comparatively large and becomes difficult as the internal friction is reduced so low that the half-width is too small. On the other hand, the decay method is more applicable when the time of half-decay is comparatively long. It becomes difficult when the internal friction increases to a value such that the time of half-decay is too short to measure accurately.

The free decay and resonance methods of measuring internal friction have been especially popular, since they require only relative, rather than absolute, measurements of amplitude of vibration. Various electrical and magnetic methods have been employed in these measurements. Wegel and Walther[11] introduced the electromagnetic method of induced eddy currents. In this method the driving end of the specimen is subjected to both a static and an oscillating magnetic field. The oscillating field induces eddy currents in the specimen, which, in turn, are pulled and pushed by the stationary, inhomogeneous, magnetic field, thereby exerting an oscillating force upon the specimen. At the detecting end the specimen is subjected likewise to a stationary magnetic field. Vibrations therefore induce eddy currents at this end. The field of these eddy currents is then detected by coils surrounding this end. This electromagnetic drive and detection through eddy currents has been used by the author in

8. F. Förster and H. Breitfeld, *Zeitschr. f. Metallk.*, XXX (1938), 343.

9. H. B. Huntington, *Phys. Rev.*, LXXII (1947), 321.

10. W. P. Mason, *Jour. A.S.A.*, XIX (1947), 464.

11. R. Wegel and H. Walther, *Physics*, VI (1935), 141.

a number of investigations.[12] A stronger coupling of the specimen with the driving and detecting system is obtained if one employs pole pieces glued onto the specimen rather than eddy currents. Such pole pieces have been employed by the author[13] in cases in which the internal friction was unusually high. An ingenious method has been adopted by Förster[14] for exciting and detecting oscillations. There, as in most methods, the specimen is slung horizontally by two wires. In most methods the two ends of each wire are attached to rigid supports. In Förster's method one end of each wire is attached to the movable membrane of an electromagnetic drive or pickup system. The vibrations are thereby excited and detected through the two supports. Still another method of exciting and detecting vibrations has been adopted by Quimby and his school.[15] In this method the specimen is cemented onto a quartz crystal, which is then excited piezoelectrically.

The measured value of the internal friction is always raised above its true value by the dissipation of energy to the surroundings. One source of energy dissipation is acoustical radiation. Such dissipation can always be reduced to negligible values by placing the specimen in an evacuated chamber. The author has found such a precaution necessary when the internal friction is as low as 10^{-4}.

As mentioned above, in most electrical methods of excitation and detection the specimen is slung horizontally on two wire supports. If the specimen suffered no motion at the supports, the dissipation of energy through the supports could be reduced essentially to zero. However, owing to the Poisson contraction, the surface of the specimen undergoes some motion even at the nodes during either transverse or longitudinal vibration. When proper precautions are taken, such dissipation of energy is, however, very small, as is evidenced by the fact that measured values of $\tan \delta$ in transverse vibrations have been as low[16] as 2×10^{-6}. In order to avoid all loss through the supports, Frommer and Murray[17] have developed special supports for use in torsional vibration.

12. R. H. Randall, F. C. Rose, and C. Zener, *Phys. Rev.*, LVI (1939), 343; C. Zener and R. H. Randall, *Trans. A.I.M.E.*, CXXXVII (1940), 41.

13. C. Zener, D. Van Winkle, and H. Nielsen, *Trans. A.I.M.E.*, CXLVII (1942), 98; and C. Zener, *Trans. A.I.M.E.*, CLII (1943), 122.

14. F. Förster, *Zeitschr. f. Metallk.*, XXIX (1937), 109.

15. S. L. Quimby, *Phys. Rev.*, XXV (1925), 558, and XXXIX (1932), 345; Zacharias, *Phys. Rev.*, XLIV (1933), 116; S. Siegel and S. L. Quimby, *Phys. Rev.*, XLIX (1936), 663; W. T. Cooke, *Phys. Rev.*, L (1936), 1158; T. A. Read, *Phys. Rev.*, LVIII (1940), 371.

16. C. Zener, H. Clarke, and C. S. Smith, *Trans. A.I.M.E.*, CXLVII (1942), 90.

17. L. Frommer and A. Murray, *Jour. Inst. Metals*, LXX (1944), 11.

In the dynamical method the internal friction of a single specimen may be measured over a range of frequencies by working at the fundamental frequency and at its first few overtones. The relation between these frequencies is trivial in the case of longitudinal and torsional vibration. The relation is given for the case of transverse vibrations in Table 8.

TABLE 8

FREQUENCIES IN TRANSVERSE VIBRATIONS*

n	ν_n/ν_0
0	1
1	2.76
2	5.41
3	8.94
4	13.37

* A. Kalähn, *Handbuch der Physik*, VIII, 201–2.

In this dynamical method of free oscillation it is usually desirable to minimize support losses by suspending the specimen at its nodes of vibration. The formula for the position of these nodes is trivial in the case of longitudinal and torsional vibration. The position of the nodes, x, in the case of transverse vibration is given by Table 9 in terms of the length of specimen, l.

TABLE 9

POSITION OF NODES IN TRANSVERSE VIBRATION (x/l)*

n	0	1	2	3	4
x/l	0.224	0.132	0.094	0.074	0.060
	0.776	.5	.356	.277	.227
	....	0.868	.644	.5	.409
	....	....	0.906	.753	.591
	....	....	....	0.927	.774
	....	....	....	....	0.940

* A. Kalähn, *Handbuch der Physik*, VIII, 201–2.

C. VARIATION OF FREQUENCY

In determining the relaxation spectrum, one must measure the internal friction over a frequency range sufficiently broad to include at least one relaxation band. As may be seen from Figure 21, the frequency range must cover at least two cycles of 10 in order that the internal friction in a band may be reduced on each side to 20 per cent of its maximum value. Such large frequency ranges have been employed.[18] Measurements over a large frequency range entail considerable experimental difficulty, and

18. K. Bennewitz and H. Rötger, *Phys. Zeitschr.*, XXXVII (1936), 578; and *Zeitschr. f. tech. Phys.*, XIX (1938), 521.

it is usually desirable to circumvent this difficulty through use of some indirect method.

As discussed in chapter v, in certain circumstances the internal friction may be obtained as a function of frequency by a single experiment upon stress relaxation at constant strain. The relation between tan δ and stress-relaxation measurements is given by equation (69). This method has been used with success by Kê[19] and is applicable whenever the relaxation band is much broader than for the case of a single relaxation time.

The stress-relaxation method has the inherent defect that determination of tan δ entails differentiation of an experimental curve, namely, the stress-relaxation curve. Differentiation of an experimentally determined curve necessarily introduces loss of accuracy as well as loss of detail. It is therefore usually desirable to employ an alternative indirect method. It is sometimes possible to predict, from theoretical considerations, that the internal friction will be a function of frequency ν and of a time of relaxation τ in the form

$$\tan \delta = f(\nu\tau) ,$$

where f is some unknown function. This function may be determined as well through a variation in τ as through a variation in ν. Thus a plot of tan δ versus log τ is identical, aside from a horizontal shift of the abscissa, to a plot of tan δ versus log ν.

The above indirect method has been used[20] to study the relaxation spectrum associated with the stress-induced intercrystalline thermal currents in polycrystalline specimens. As discussed in Section A-4 of chapter vii, the time of relaxation τ may in this case be set equal to d^2/D, where d is the mean grain diameter and D is the thermal diffusion coefficient. An effective frequency range covering five cycles of 10 was obtained by varying d, i.e., by examining specimens of different mean grain diameter.

This indirect method has also been applied[21] to several cases in which the time of relaxation τ is suspected to vary with temperature according to an Arrhenius equation, namely, as

$$\tau \sim e^{-H/RT} .$$

One thereby obtains a plot of tan δ versus $ln\ \nu$ by making measurements at one frequency over a range of temperatures and by plotting the results versus (H/RT). The heat of activation H is obtained by making observa-

19. T. S. Kê, *Phys. Rev.*, LXXI (1947), 533.

20. R. H. Randall, F. C. Rose, and C. Zener, *Phys. Rev.*, LVI (1939), 343.

21. J. L. Snoek, *Physica*, IX (1942), 862; C. Zener, *Trans. A.I.M.E.*, CLII (1943), 122, and *Phys. Rev.*, LXXI (1947), 34; L. Dijkstra, *Philips Research Repts.*, II (1947), 357; T. S. Kê, *Phys. Rev.*, LXXI (1947), 533, and LXXII (1947), 41.

tions at two frequencies. Thus, suppose tan δ versus $1/T$ curves are obtained at the two frequencies ν_1 and ν_2. Then, if tan δ is indeed a function of the product $\nu\tau$ where τ obeys an Arrhenius equation, these two curves may be superimposed by a horizontal shift. If $\Delta(1/T)$ is the magnitude of this shift, then the heat of activation H is determined from the relation

$$ln\,(f_2/f_1) = (H/R)\cdot\Delta(1/T)\,.$$

As an example, an effective frequency range of three cycles of 10 will be traversed in raising the temperature from 0° to 60° C. when the heat of activation is 20,000 cal/gm/mole.

VII

PHYSICAL INTERPRETATION OF ANELASTICITY

A. HOMOGENEOUS RELAXATION

IN NO real metal is the strain in the preplastic range a function of stress alone. It changes also with temperature, with composition, and in some metals with magnetic field or perhaps degree of order. In nonconducting solids the strain may also be changed by an electric field. The dependence of strain upon thermodynamically specifiable variables other than stress is a common source of anelasticity.

As an introduction to a general discussion, the well-known example of thermoelastic coupling will be reviewed in detail. The results of this review will be applicable to thermodynamical variables other than temperature.

For the present we shall consider that the strain depends only upon the two variables, stress and temperature. The strain, referred to zero stress and to a standard temperature as the reference configuration, will then be given by

$$e = M_R^{-1}\sigma + \lambda \Delta T \,. \qquad (89)$$

In this equation the relaxed modulus, M_R, is also known as the "isothermal modulus," the coefficient λ is the linear thermal expansion coefficient, and ΔT is the deviation of the temperature from the standard temperature. Obviously, a relation cannot be obtained between stress and strain unless the variation in temperature is specified by an independent equation. Now two phenomena induce temperature changes. The first of these is diffusion, that is, the equalization, or *relaxation*, of temperature fluctuations. In many important cases, discussed in detail below, the change in temperature due to diffusion may be expressed to a very good approximation by the following simple relaxation equation:

$$\left(\frac{d}{dt}\Delta T\right)_{\text{diffusion}} = -\tau^{-1}\Delta T \,. \qquad (90)$$

The quantity τ will be known as the "relaxation time." The precise value of the relaxation time will depend upon the restrictions placed upon stress or strain during relaxation. If the stress is maintained constant and the strain relaxes with temperature, we shall have one relaxation time. On the other hand, if the strain is maintained constant and the stress relaxes with temperature, we shall have another relaxation time. These two values will be denoted by τ_σ and τ_e, respectively. The second phenomenon that affects temperature is a change in strain which arises from sources other than the temperature change itself. Thus, if a rise in temperature induces an increase in length, conversely an adiabatic increase in length must be associated with a decrease in temperature. This second change in temperature may therefore be written as

$$\left(\frac{d}{dt}\Delta T\right)_{\text{adiabatic}} = -\gamma \dot{e}\,, \tag{91}$$

where

$$\gamma = \left(\frac{\partial T}{\partial e}\right)_{\text{adiabatic}}.$$

Upon combining equations (90) and (91), we obtain

$$\frac{d}{dt}\Delta T = -\tau_e^{-1}\Delta T - \gamma \dot{e}\,, \tag{92}$$

where, in accordance with our prior definition of the two types of relaxation time, τ_e denotes the relaxation time for constant strain.

An equation governing the relation between stress and strain is now derived by eliminating ΔT between equations (89) and (92). One thus obtains

$$M_R e + M_U \tau_e \dot{e} = \sigma + \tau_e \dot{\sigma}\,, \tag{93}$$

where the quantity M_U is defined by

$$M_U = (1 + \lambda\gamma) M_R \tag{94}$$

and, from the form of equation (93), denotes the unrelaxed modulus. One may therefore interpret τ_e as the relaxation time for stress relaxation, as well as for temperature relaxation, at constant strain. It therefore seems that τ_σ may be the relaxation time for strain relaxation, as well as for temperature relaxation at constant stress. If this is true, then equation (93) may be written as

$$M_R(e + \tau_\sigma \dot{e}) = \sigma + \tau_e \dot{\sigma}\,, \tag{95}$$

and the relation between the two times of relaxation is

$$\frac{\tau_\sigma}{\tau_e} = \frac{M_U}{M_R}.$$

The identity of the τ_σ defined in this manner with the original definition of relaxation time for temperature at constant stress may be verified by the elimination of e and $\dot{e}$ between equations (89) and (95), leading to

$$\frac{d}{dt}\Delta T = -\tau_\sigma^{-1}\Delta T - \gamma M_U \dot{\sigma}\,. \tag{96}$$

We conclude that, whenever the temperature fluctuation is governed by equation (92), the thermoelastic coupling changes the solid from a Hooke solid into a standard linear solid, whose anelastic properties were discussed in detail in chapter v, section A. Thus the dependence of internal friction and elastic modulus upon frequency is given by Figure 21, while the relaxation of stress at constant strain, or of strain at constant stress, is given by Figures 18–20.

Considerable physical insight into the reason why the internal friction has a maximum in the vicinity of the angular frequency $\omega \simeq 1/\tau$ and approaches zero asymptotically on either side of this maximum may be gained from an analysis of the precise mechanism whereby energy is dissipated as heat. If a quantity of heat, δQ, is added to a unit volume at temperature T, the entropy, S, of the unit volume increases by the amount

$$\delta S = \frac{\delta Q}{T} .$$

The total increase in entropy during a cycle of vibration is therefore

$$\Delta S = \Sigma \frac{\delta Q}{T} , \tag{97}$$

where the summation extends over all increments during one cycle of vibration. Once a steady state is reached, the specimen suffers no net increase in entropy, the net flow of heat away from the specimen exactly balancing the dissipation of energy within the specimen. If T_0 is the mean temperature of the specimen, and δT is its deviation from the mean temperature at any one instant, then the equilibrium condition,

$$\Delta S = 0 ,$$

may be re-written, with the help of equation (97), as

$$\Delta Q = -T_0^{-1} \cdot \Sigma \, \delta Q \cdot \delta T \tag{98}$$

where the net outflow of heat per cycle is

$$\Delta Q = \Sigma \, \delta Q .$$

From equation (98) we see that the energy dissipated per cycle contains a summation of products of the type $\delta Q \cdot \delta T$. In the very high frequency range, where the vibration is essentially adiabatic, the energy dissipated is very low because of the smallness of the first factor, δQ. In the very low frequency range, where the vibration is essentially isothermal, the energy dissipated is very low because of the smallness of the second factor, δT. Only in the intermediate frequency range, where the vibration is neither essentially adiabatic nor isothermal and where both δQ and δT are sensibly different from zero, is the dissipation of energy appreciable.

From equation (94) we obtain the following formula for the relaxation strength, as defined in equation (71):

$$\Delta = \lambda\gamma \,. \tag{99}$$

An alternative expression will be derived in terms of quantities which are more directly measurable than γ. If we regard σ and T as the independent variables and e and S as the dependent variables, then

$$de = M_T^{-1} d\sigma + \lambda dT \,, \tag{100}$$

$$dS = \lambda d\sigma + \mu dT \,, \tag{101}$$

where

$$\mu = \left(\frac{\partial S}{\partial T}\right)_\sigma . \tag{102}$$

The symmetry of the coefficients follows from the fact that the increment of the free energy density is a perfect differential. Thus from

$$d(E - \sigma e - TS) = -e d\sigma - SdT \,, \tag{103}$$

it follows that

$$\left(\frac{\partial e}{\partial T}\right)_\sigma = \left(\frac{\partial S}{\partial \sigma}\right)_T . \tag{104}$$

Upon setting dS equal to zero in equation (101), we obtain for M_S^{-1}

$$\left(\frac{\partial e}{\partial \sigma}\right)_S = M_T^{-1} - \frac{\lambda^2}{\mu} ,$$

and therefore

$$\Delta = \frac{M_S \lambda^2}{\mu} . \tag{105}$$

Still another expression for the relaxation strength may be derived by observing from equations (100) and (101) that

$$\left(\frac{\partial e}{\partial S}\right)_\sigma = \frac{\lambda}{\mu} .$$

Substitution of this relation in equation (105) leads to

$$\Delta = M_S \mu \left(\frac{\partial e}{\partial S}\right)_\sigma^2 . \tag{105a}$$

I. THERMODYNAMICAL VARIABLES

As previously intimated, temperature is only one of several thermodynamical variables in which fluctuations are introduced by strains and for which the relaxation of such fluctuations gives rise to anelasticity. In the present section the various types of thermodynamical variables are discussed. In the following sections the various mechanisms of relaxation are analyzed.

As a guide in the formulation of a general theory of relaxation, we shall apply thermodynamical considerations. If a system of unit volume is completely specified by stress and temperature, the increase in energy associated with a change in state is given by

$$dE = \sigma d e + TdS , \tag{106}$$

where S is the entropy of the system, i.e., the entropy of an amount of material which has unit volume at zero stress and at a standard temperature. Now T may be regarded as a thermodynamical potential, since a system is in equilibrium only when everywhere T has the same value. On the other hand, the entropy per unit volume may be regarded as a thermodynamical density, since the entropy of a system is equal to the entropy of its component parts. Now when more variables are necessary completely to specify the system, it is possible to generalize equation (106) to the form

$$dE = \sigma d e + \Sigma_j A_j d\alpha_j , \tag{107}$$

where all the A's may be regarded as thermodynamical potentials and the α's as thermodynamical densities. Equation (89) may then be generalized to

$$e = M_R^{-1} \sigma + \Sigma_j \lambda_j \Delta A_j . \tag{108}$$

In this equation each ΔA_j is the deviation of the corresponding potential from a standard value, e is the strain referred to the standard state in which $\sigma = 0$ and all the potentials have their standard values, and the coefficient λ_j is defined as

$$\lambda_j = \frac{\partial e}{\partial A_j}, \tag{109}$$

the partial derivative denoting that the stress and all the potentials other than A_j are to be maintained constant during the differentiation.

In many cases thermodynamical potentials other than temperature obey a relaxation equation such as equation (92), i.e.,

$$\frac{d\Delta A_j}{dt} = - \tau_j^{-1} \Delta A_j - \gamma_j \dot{e} . \tag{110}$$

The relaxation of such thermodynamical potentials contributes to the anelasticity of the specimen in precisely the same manner as does the relaxation of temperature fluctuations. In particular, the relative difference between the associated relaxed and unrelaxed moduli is given by

$$\frac{M_{a_j} - M_{A_j}}{M_{A_j}} = \lambda_j \gamma_j , \qquad (111)$$

or, in analogy to equation (105), by

$$\frac{M_{a_j} - M_{A_j}}{M_{A_j}} = \frac{M_{a_j} \lambda_j^2}{\mu_j} , \qquad (111a)$$

where

$$\mu_j = \left(\frac{\partial a_j}{\partial A_j} \right)_{\text{other } A\text{'s}, \sigma} . \qquad (111b)$$

In deriving this equation it has been implicitly assumed that the same thermodynamical potentials or densities, other than A_j or a_j, are held constant in the measurement of M_{a_j} as in M_{A_j}.

In a specimen which contains a mixture of elements all in solid solution, strain induces changes in the partial thermodynamical potentials of each constituent. These partial thermodynamical potentials, sometimes called "chemical potentials," will be referred to unit volume rather than to a gram-mole, as is more usual. They are defined as follows: Let G be the Gibbs free energy of an amount of material which contains unit volume under standard conditions, i.e., at zero stress and at standard composition. Thus

$$G = E - e\sigma - TS . \qquad (112)$$

Then the partial thermodynamical potential, G_j, will be defined as

$$G_j = \frac{\partial G}{\partial n_j} , \qquad (113)$$

where n_j is the number of atoms of type j and where, in the differentiation, σ, T, and the number of all other types of constituent atoms are to be maintained constant. Upon solving equation (112) for E and differentiating, we find, therefore,

$$dE = \sigma d e + T dS + \Sigma G_j dn_j . \qquad (114)$$

The G_j's and the deviation of the n_j's from the standard concentrations may therefore be regarded as thermodynamical potentials and densities, respectively. The G_j at each part of a specimen will be changed by strain. Since equilibrium is attained only when the G_j's are constant throughout the specimen, a fluctuation of the G_j's will gradually be equalized by diffusion. Thus relaxation of the fluctuations in the G_j's will be accompanied by the segregation into the regions under extension of those atoms which increase the lattice constant and, conversely, the segregation into those regions under compression of those atoms which decrease the lattice constant.

In ferromagnetic materials the coupling between strain and external magnetic field is well known. In many cases of importance the external magnetic field is essentially linear, and the axis of the specimen is parallel to this field. If the transverse dimensions of the specimen are very small compared with the wave length of the strain within the specimen, then the macroscopic magnetic field within the specimen may be regarded everywhere as parallel to the axis of the specimen. Application of the Maxwell equation,

$$\operatorname{curl} H = 0.4\pi i + c^{-1}\frac{dD}{dt}, \qquad (115)$$

to this case shows that under equilibrium conditions, when i and dD/dt are both zero, the macroscopic magnetic field strength, H, is constant across the specimen and is equal to the external magnetic field strength. The macroscopic magnetic field strength, H, may therefore, under the above conditions, be regarded as a thermodynamical potential. In order to find the associated thermodynamical density, we note that changes in the magnetic energy density are given by

$$(dE)_{\text{magnetic}} = \frac{1}{4\pi} H dB ,$$

where B is the magnetic flux density. Thus, if H is regarded as a thermodynamical potential, $B/4\pi$ must be regarded as the associated density. The relaxation of the magnetic strength fluctuations induced by strains is one important cause of anelasticity.

One cannot, strictly, speak of the "temperature" of a specimen unless all parts of the specimen are in thermal equilibrium, i.e., unless the temperature is uniform throughout. Likewise, one cannot even speak of the temperature of an element of volume unless all degrees of freedom within the element are in thermal equilibrium. In the previous discussion of temperature relaxation and of the anelasticity associated therewith, it was implicitly assumed that all degrees of freedom within each element of volume were in thermal equilibrium and hence that a temperature could be associated with each element. In many important cases such an assumption is not valid. Rather it is necessary to divide the normal coordinates into two groups, the normal coordinates within the same group being in thermal equilibrium. Anelasticity then arises through the relaxation of the temperature difference between the two groups of coordinates. One group contains the coordinates associated with thermal vibration of the atoms. The other coordinates may determine the distribution of atoms on the lattice positions in case more than one type of atom is present. Such coordinates are known as "ordering" coordinates. In the case of molecular crystals the second group of coordinates may refer to the coordinates of the molecules with respect to their center of gravity. With each group of coordinates there is associated a corresponding entropy density.

II. ORTHOGONAL THERMODYNAMIC POTENTIALS

In the above discussion of the anelasticity associated with relaxation, it was explicitly assumed that the thermodynamic potentials satisfied a simple relaxation equation of the type of equation (110). Such an assumption is probably valid for the relaxation associated with the establishment of thermal equilibrium between the various coordinates of an element of volume or for the relaxation of a temperature fluctuation between a specimen and the surrounding medium when the specimen may be regarded as at a uniform temperature throughout. In those cases, however, in which the relaxation relates to fluctuations in thermodynamic potentials between different regions of the specimen, the simple relaxation equation cannot be applied without further modification.

One possible modification is to replace the thermodynamic potentials, which are functions of position, with an infinite series of normalized orthogonal functions, the latter satisfying the appropriate diffusion equation and boundary conditions. This procedure, originally developed by the author,[1] is described below. An alternative procedure has been described by Päsler.[2] The coefficient of each of the above normal orthogonal functions is a function of time and satisfies a simple relaxation equation of the type of equation (110). These coefficients will have the dimensions of thermodynamic potentials and will therefore be called "orthogonal thermodynamic potentials." Each type of relaxation, be it thermal, concentration, or magnetic, will then involve the simultaneous relaxation of all associated orthogonal thermodynamic potentials. When the total relative relaxation is small compared with unity, i.e., when

$$\Delta \ll 1 \ ,$$

the internal friction may be expressed as the sum of the contributions from the individual potentials. As an introduction to the general analysis of this method, an example of thermal relaxation will be presented.

1. C. Zener, "General Theory of Thermoelastic Internal Friction," *Phys. Rev.*, LIII (1938), 90.

2. M. Päsler, "Zur Theorie der thermische Dämpfung in fester Körper," *Zeitschr. f. Phys.*, CXXII (1944), 357.

For this example we shall consider the internal friction arising from relaxation of temperature across a specimen vibrating transversely. The two equations which must be solved simultaneously are

$$e = M_R^{-1}\sigma + \lambda\Delta T \tag{116}$$

and

$$\frac{d\Delta T}{dt} = D\nabla^2\Delta T - \gamma \dot{e}\,. \tag{117}$$

The second equation replaces the simple relaxation relation of equation (92). In order to derive equations in which the spatial coordinates are eliminated, we expand e, σ, and ΔT in a complete set of normal orthogonal functions. Thus,

$$e = \Sigma_j e_j(t)\, U_j\,, \tag{118}$$

$$\sigma = \Sigma_j \sigma_j(t)\, U_j\,, \tag{119}$$

$$\Delta T = \Sigma_j T_j(t)\, U_j\,. \tag{120}$$

These functions will be chosen as the characteristic solutions of the differential equation

$$D\nabla^2 U + \tau^{-1} U = 0 \tag{121}$$

and of an appropriate boundary condition. Such solutions exist only when τ has discrete values, known as the "characteristic relaxation times."

In the particular case under consideration, strain may be regarded, to a close approximation, as a function only of the transverse dimension of the specimen. Equation (121) may therefore be replaced by

$$\left(\frac{Dd^2}{dx^2} + \tau^{-1}\right) U = 0\,, \tag{122}$$

the coordinate x being taken as this transverse dimension. The boundary condition will be taken to correspond to no heat flow across the boundaries. If the transverse width of the specimen is a, the boundary condition will then be

$$\frac{dU}{dx} = 0 \text{ at } X = \pm\frac{a}{2}.$$

The characteristic normalized solutions and relaxation times for this problem are

$$U_k = \sqrt{\frac{2}{a}} \sin\left\{(2k+1)\frac{\pi x}{a}\right\}$$

and

$$\frac{1}{\tau_k} = (2k+1)^2 \frac{\pi^2 D}{a^2}\,,$$

respectively, where k is any one of the sequence of numbers 0, 1, 2, 3, ...

If we now multiply equations (116) and (117) by U_j, integrate over the volume of the specimen, and make use of equation (121), we obtain the following two equations:

$$e_k = M_R^{-1}\sigma_k + \lambda T_k \tag{123}$$

and

$$\dot{T}_k = -\tau_k^{-1} T_k - \gamma \dot{e}_k \, . \tag{124}$$

Each orthogonal thermodynamic potential, T_k ,therefore satisfies a simple relaxation equation.

Now the rate of dissipation of energy by the specimen is given by

$$\dot{E} = \int \sigma \dot{e} d v \, , \tag{125}$$

the integral extending throughout the volume of the specimen. Upon making use of equations (118) and (119), we arrive at

$$\dot{E} = \Sigma_j \sigma_j \dot{e}_j \, . \tag{126}$$

If we now suppose that e, σ, and T vary periodically with time and if we take the time average, we obtain, upon using equation (123),

$$\overline{\dot{E}} = -M_R \lambda \Sigma_j \overline{\dot{e}_j T_j} \, . \tag{127}$$

Taking ω as the angular frequency of vibration, we derive from equation (124)

$$\overline{\dot{e}_j T_j} = -\frac{\gamma \omega^2 \tau}{1 + \omega^2 \tau^2} \overline{e_j^2} \, . \tag{128}$$

In the case in which M_R/M_U is only slightly less than unity, the internal friction may be written as in equation (80), where ΔE is the energy loss per cycle and E is the average energy of vibration. Now

$$\Delta E = \frac{2\pi}{\omega} \overline{\dot{E}} \, , \tag{129}$$

and, when M_R/M_U is only slightly less than unity,

$$E = M_R \int \overline{e^2 d v} = M_R \Sigma_j \overline{e_j^2} \, . \tag{130}$$

Upon combining equations (127)—(130), we obtain

$$\tan \delta = \lambda \gamma \left(\frac{\Sigma_j \overline{e_j^2} \dfrac{\omega \tau_j}{1 + \omega^2 \tau_j^2}}{\Sigma \overline{e_j^2}} \right) . \tag{131}$$

If we now make use of equation (94) and define the quantity

$$f_j = \frac{\overline{e_j^2}}{\Sigma_j \overline{e_j^2}}, \tag{132}$$

we transform equation (131) into

$$\tan \delta = \Delta_E \Sigma_j f_j \frac{\omega \tau_j}{1 + \omega^2 \tau_j^2}. \tag{133}$$

By the manner in which they are defined, the quantities f_j automatically satisfy the following condition:

$$\Sigma f_j = 1 . \tag{134}$$

The evaluation of internal friction due to relaxation may thus be reduced to the problem of finding the characteristic relaxation times, τ_j, and the coefficients, f_j. The characteristic relaxation times have already been found for the special case under consideration. The coefficient f_j is given by

$$f_j = \frac{\left\{\left(\frac{2}{a}\right)^{1/2} \int_{-a/2}^{a/2} x \sin (2j+1)(\pi x/a)\, dx\right\}^2}{\int_{-a/2}^{a/2} x^2 dx} \tag{135}$$

$$= 96\pi^{-4}(2j+1)^{-4} .$$

The first few values of the coefficients are given by

$$f_0 = 0.986,\ f_1 = 0.012,\ f_2 = 0.0016 .$$

In this particular case, only a very slight error would be made by setting f_0 equal to unity and all the following f's equal to zero. The thermal relaxation across a specimen vibrating transversely may therefore be regarded as essentially a simple relaxation with

$$\text{Relaxation time} = a^2/\pi^2 D;\ a = \text{Width}.$$

III. GENERAL THEORY OF RELAXATION BY DIFFUSION

In the preceding section the concept of orthogonal thermodynamic potentials was introduced, and it was shown how the internal friction arising from diffusion could be computed by use of such potentials. In the present section the results of such computations in particular cases are summarized.

It has been shown that each orthogonal thermodynamic potential relaxes in a simple manner and therefore contributes an internal friction which varies with frequency in the manner shown in Figure 21. The contribution from all the orthogonal potentials may therefore be expressed as in equation (133), namely,

$$\tan \delta = \Delta_E \cdot \Sigma_n f_n \frac{\omega_n \tau_n}{1 + \omega_n^2 \tau_n^2}. \tag{136}$$

Each coefficient f_n may be regarded as the relative strength of the associated orthogonal potential. The sum of all the relative strengths is unity, as indicated by equation (134). The relative strengths are computed as follows: Let e_0 (x, y, z) be the strain when the amplitude of the vibration is at a maximum. Further, let U_n (x, y, z) be the nth normalized orthogonal potential. Then

$$f_n = \frac{(\int U_n e_0 d v)^2}{\int e_0^2 d v} . \tag{137}$$

The orthogonal potentials are the characteristic solutions of the equation

$$(D\nabla^2 + \tau^{-1})\, U = 0 \tag{138}$$

and of the associated boundary conditions. The characteristic relaxation times, τ_n, are those values of τ for which such solutions exist.

In one important class of diffusion problems all terms in the summation in equation (136) are negligible compared with the first. Such was the case in the example cited in the previous section. In this class of problems the diffusion is essentially a volume diffusion, there being no diffusion across surfaces at which the properties of the medium suffer a discontinuous change.

In a second important class of problems, successive terms in the summation in equation (136) converge very slowly, particularly at high frequencies. In such problems the slow convergence is due to the presence of surfaces across which diffusion takes place and across which the physical properties of the medium change discontinuously. In the absence of all diffusion the thermodynamic potentials would suffer a discontinuous change across these surfaces during vibration. Therefore, at increasingly high frequencies, where diffusion per half-cycle becomes less and less, the gradient of the potentials across these surfaces becomes steeper and steeper. At increasingly high frequencies the diffusion therefore becomes limited essentially to the immediate vicinity of these surfaces and is normal thereto. Two alternative methods may be used for computing the internal friction in those cases of high frequency when the convergence of the summation in equation (136) is poor. One may convert the summation into an integral, thus

$$\tan\ \delta = \Delta_E \int f_n \frac{\omega_n \tau_n}{1 + \omega_n^2 \tau_n^2}\, dn . \tag{139}$$

On the other hand, one may take cognizance of the fact that under these conditions of high frequency the diffusion is essentially confined to the discontinuous surfaces and must therefore be proportional to the area of such surfaces per unit volume. Thus the only computation required is the evaluation of the dissipation of energy per unit area of such surfaces. This computation is given in the appendix to this section. The result is

$$\tan \delta = \Delta_E \left(\frac{D}{2\omega}\right)^{1/2} \left(\frac{\text{area}}{\text{volume}}\right) \cdot \beta . \tag{140}$$

The precise value of the numerical coefficient β depends upon the type of system. If the only surface is the boundary of the specimen, then

$$\beta = \frac{(e_0^2) \text{ surface average}}{(e_0^2) \text{ volume average}} . \tag{141}$$

If, on the other hand, internal boundaries are present across which properties change discontinuously, such as grain boundaries, the evaluation of β requires special consideration. A particular case is examined below.

a) DIFFUSION BETWEEN SPECIMEN AND SURROUNDING MEDIUM

i. *Longitudinal vibration of specimen of circular cross-section.*[3]—The relative strength, f_n, is given by

$$f_n = \frac{4}{q_n^2} ,$$

where q_n is the nth root of

$$J_0(q) = 0 .$$

The characteristic relaxation time, τ_n, is given by

$$\tau_n = \frac{a^2}{D q_n^2} ,$$

where a is the radius of the specimen. From the formula[4]

$$\Sigma q_n^{-4} = \frac{1}{32} ,$$

one obtains the asymptotic relation,

$$\tan \delta = \Delta_E \cdot \frac{a^2 \omega}{D} , \qquad \omega \ll \frac{D}{a^2} , \tag{142}$$

while from equations (140) and (141) one obtains the asymptotic relation,

$$\tan \delta = \Delta_E \cdot \left(\frac{2D}{a^2 \omega}\right)^{1/2} , \qquad \omega \gg \frac{D}{a^2} . \tag{143}$$

These asymptotic relations were originally derived by Kersten.[5]

ii. *Longitudinal vibration of reeds.*—The relative strength is given by

$$f_n = \frac{8}{(2n+1)^2 \pi^2} .$$

3. C. Zener, "General Theory of Macroscopic Eddy Currents," *Phys. Rev.*, LIII (1938), 1010.

4. G. N. Watson, *Theory of Bessel Functions* (Cambridge, 1922), p. 502.

5. M. Kersten, *Zeitschr. f. tech. Phys.*, XV (1934), 463.

The characteristic relaxation time, τ_n, is given by

$$\tau_n = \frac{4a^2}{(2n+1)^2\pi^2 D},$$

where a is the transverse width of the reed.

From the formula[6]

$$\Sigma(2n+1)^{-4} = \frac{\pi^4}{96},$$

one obtains the asymptotic relation,

$$\tan\,\delta = \Delta_E \cdot \frac{a^2\omega}{3D}, \qquad \omega \ll \frac{D}{a^2}, \qquad (144)$$

while from equations (140) and (141) one obtains

$$\tan\,\delta = \Delta_E\left(\frac{D}{2a^2\omega}\right)^{1/2}, \qquad \omega \gg \frac{D}{a^2}. \qquad (145)$$

iii. *Transverse vibration of specimen with circular cross-section.*—The relative strength, f_n, is given by

$$f_n = \frac{8}{q_n^2},$$

where q_n is the nth root of

$$J_1(q) = 0\,.$$

The characteristic relaxation time, τ_n, is given by

$$\tau_n = \frac{a^2}{q_n^2 D}.$$

From the formula[7]

$$\Sigma q_n^{-4} = \frac{1}{192},$$

one obtains the asymptotic relation,

$$\tan\,\delta = \Delta_E \frac{a^2\omega}{24D}, \qquad \omega \ll \frac{D}{a^2}, \qquad (146)$$

while from equations (140) and (141) one finds

$$\tan\,\delta = \Delta_E\left(\frac{8D}{a^2\omega}\right)^{1/2}, \qquad \omega \gg \frac{D}{a^2}. \qquad (147)$$

iv. *Transverse vibration of reeds.*—The relative strength, f_n, is given by

$$f_n = \frac{6}{\pi^2 n^2},$$

6. E. P. Adams, *Smithsonian Mathematical Formulae* (Washington, D.C.: Smithsonian Institution, 1922), p. 140.

7. Watson, *op. cit.*, p. 502.

while the characteristic relaxation time, τ_n, is given by

$$\tau_n = \frac{a^2}{\pi^2 n^2 D},$$

where $2a$ is the transverse width.

From the formula[8]

$$\Sigma n^{-4} = \frac{\pi^4}{90},$$

one obtains the asymptotic relation,

$$\tan\ \delta = \Delta_E \cdot \frac{a^2 \omega}{15 D}, \qquad \omega \ll \frac{D}{a^2}. \qquad (148)$$

From equations (140) and (141) one finds the asymptotic relation

$$\tan\ \delta = \Delta_E \cdot 3 \left(\frac{D}{2a^2 \omega}\right)^{1/2}, \qquad \omega \gg \frac{D}{a^2}. \qquad (149)$$

b) MACROSCOPIC DIFFUSION WITHIN SPECIMEN

When diffusion is confined to the interior of the specimen and when the diffusion is essentially macroscopic in nature, the following summary shows that no appreciable error is made by retaining only the first term in the infinite summation in equation (136). The thermodynamic potential giving rise to diffusion may thus be regarded as obeying a simple relaxation equation. As we have previously seen, the analysis for the coupling between strain and stress, on the one hand, and the thermodynamical potential, on the other hand, may in this case be carried out exactly, irrespective of the magnitude of M_U/M_R. According to equation (56), the internal friction may therefore be written as

$$\tan\ \delta = \frac{M_U - M_R}{(M_U M_R)^{1/2}} \frac{\omega \tau}{1 + \omega^2 \tau^2},$$

where

$$\tau = (\tau_\sigma \tau_e)^{1/2}.$$

On the other hand, the elastic aftereffects may be obtained precisely through the solution of equation (93) or equation (95). Therefore the relaxation of stress following sudden application of a constant strain is given by equation (46), the relaxation of strain following the sudden application of a constant stress is given by equation (48).

i. *Longitudinal vibration.*—$f_0 = 1$, precisely.

$$\tau_0 = \frac{D}{C^2},$$

where C is the velocity of propagation.

8. Adams, *op. cit.*

ii. *Transverse vibration of reeds.*[9]—The relative strengths are given by

$$f_n = 96\pi^{-4}(2n+1)^{-4} ,$$

the first few values of which are 0.986, 0.012, 0.0016. The longest characteristic relaxation time is given by

$$\tau_0 = a^2\pi^{-2}D^{-1} ,$$

where a is the transverse width; successive values are given by

$$\tau_n = \frac{\tau_0}{(2n+1)^2} .$$

iii. *Transverse vibration of specimens with circular cross-section.*[10]—The relative strengths are given by

$$f_n = 8q_n^{-2}(q_n^2 - 1)^{-1} ,$$

where q_n is the nth root of

$$J_1^1(q) = 0 .$$

The first few values are 0.988, 0.010, and 0.0015. The characteristic relaxation times are given by

$$\tau_n = \frac{a^2}{Dq_n^2} ,$$

the first few values of which are 0.295 a^2D, 0.035 a^2D, and 0.014 a^2D.

c) INTERGRAIN DIFFUSION

In polycrystalline specimens of noncubic metals the elastic anisotropy and at least partial random orientation result in a fluctuation of stress and strain from grain to grain. Application of a stress will therefore result in a fluctuation of the thermodynamic potentials from grain to grain, a fluctuation which is relaxed by intergrain diffusion. Such intergrain diffusion will give rise to characteristic anelastic effects.

No exact computation can be made for the difference between the relaxed and the unrelaxed moduli associated with intergrain diffusion. At best, an estimate may be obtained. Suppose, for this purpose, that the stress in each grain is uniform and is equal to the stress in every other grain and is a uniaxial tensile stress. The rise in thermodynamic potential due to the adiabatic application of this stress is, in every grain, equal to

$$\Delta A = \left(\frac{\partial A}{\partial \sigma}\right)_a \sigma .$$

9. C. Zener, "Theory of Internal Friction in Reeds," *Phys. Rev.*, LII (1937), 230.

10. C. Zener, "General Theory of Thermoelastic Internal Friction," *Phys. Rev.*, LIII (1938), 90.

From the equations

$$de = M_R^{-1} d\sigma + \lambda dA ,$$

$$d\alpha = \lambda d\sigma + \mu dA ,$$

which are analogous to equations (100) and (101), we obtain

$$\left(\frac{\partial A}{\partial \sigma}\right)_\alpha = -\frac{\lambda}{\mu};$$

and therefore

$$\Delta A = -\frac{\lambda}{\mu}\sigma .$$

The coefficient λ will be dependent upon crystallographic orientation in noncubic metals. Therefore ΔA will vary from grain to grain. The deviation of ΔA from the mean is, according to the above assumption of uniform stress, given by

$$\delta A = -(\lambda - \bar{\lambda})\mu^{-1}\sigma .$$

We now compute the strain in each grain which will result from the equalization of potential fluctuations. The strain in each grain will be given by

$$\Delta e = -\left(\frac{\partial e}{\partial A}\right)_\sigma \delta A = \lambda(\lambda - \bar{\lambda})\mu^{-1}\sigma .$$

The over-all strain resulting from the equalization of potential differences between adjacent grains will be the average of this strain, namely,

$$\overline{\Delta e} = (\overline{\lambda^2} - \bar{\lambda}^2)\mu^{-1}\sigma .$$

The change in the modulus due to local relaxation between grains may therefore be written as

$$M_R^{-1} - M_U^{-1} = \mu^{-1}\overline{\lambda^2}\cdot R \tag{150}$$

where the anisotropy factor R is defined by

$$R = \frac{\overline{\lambda^2} - \bar{\lambda}^2}{\overline{\lambda^2}} . \tag{151}$$

The first factor in equation (150) is the change in M^{-1} which would result from complete relaxation of all changes in the thermodynamic potential. The second factor, R, is that fraction of this change which takes place by intergrain diffusion, that is, by relaxation of the fluctuations in the potential which occur from grain to grain.

In metals of trigonal and hexagonal symmetry λ varies with orientation as

$$\lambda = \lambda_{11}\cos^2\theta + \lambda_{\perp}\sin^2\theta . \tag{152}$$

Substitution into equation (151) leads to

$$R = \tfrac{4}{3}(\lambda_{11} - \lambda_{\perp})^2 (3\lambda_{11}^2 + 4\lambda_{11}\lambda_{\perp} + 8\lambda_{\perp}^2)^{-1} .$$

A similar analysis for polycrystalline specimens of cubic metals gives a zero fluctuation of thermodynamic potentials between adjacent grains, since in cubic metals the coefficient λ is independent of crystallographic orientation. A more detailed analysis, in which we no longer assume a uniform stress, shows that even in cubic metals a sudden application of stress will result in fluctuations of thermodynamic potentials between adjacent grains, the relaxation of which will give rise to anelastic effects. Thus the constraints imposed by neighboring grains prevent uniaxiality of stress within the grains. As an example, consider a grain boundary normal to the tensile axis. The Poisson contraction normal to this axis will be different for the two grains on either side of the boundary. At the boundary, however, they must contract equally. This boundary constraint is accompanied by transverse stresses at and in the vicinity of the boundaries. The inequality of these transverse stresses results in an inequality in the thermodynamic potentials and hence in diffusion accompanied by anelastic effects. Any estimate of the magnitude of the difference between the relaxed and the unrelaxed moduli will lead to an expression of the type in equation (150), where now the factor R is the relative mean square deviation of some function of the elastic constants. For ease of computation we shall take the following:

$$R \simeq \frac{\{(E^{-1} - (E^{-1})_{\text{av}})^2\}_{\text{av}}}{(E^{-2})_{\text{av}}} . \qquad (153)$$

Upon using the relation[11]

$$E^{-1} = s_{11} - 2 \left\{ (s_{11} - s_{12}) - \frac{s_{44}}{2} \right\} (\gamma_1^2\gamma_2^2 + \gamma_2^2\gamma_3^2 + \gamma_3^2\gamma_1^2) ,$$

where γ_1, γ_2, and γ_3 refer to the three direction cosines of the axis of the specimen with respect to the crystallographic axes, one obtains

$$R = \tfrac{4}{525} q^2 \{1 - \tfrac{2}{5} q + \tfrac{1}{21} q^2\}^{-1} , \qquad (154)$$

where the quantity

$$q = 2 \frac{(s_{11} - s_{12}) - s_{44}}{s_{11}} \qquad (155)$$

is a measure of the elastic anisotropy of the grains. The quantities q and R for a number of cubic metals are tabulated in Table 10.

Irrespective of whether the polycrystalline specimen is of a noncubic

11. E. Schmidt and W. Boas, *Kristallplastizität* (Berlin: Springer, 1935), p. 23.

or of a cubic metal, at sufficiently high frequencies diffusion will be confined essentially to the immediate vicinity of the interfaces between grains, across which the elastic constants change discontinuously. In this case an approximation to the internal friction arising from this diffusion may be obtained by use of equation (140). The grain-boundary area per unit volume is inversely proportional to the mean linear dimensions of the grains. We thus obtain

$$\tan\,\delta \simeq \Delta \frac{\left(\frac{D}{2\omega}\right)^{1/2}}{\text{G.S.}}, \tag{156}$$

where G.S. denotes any reasonable measure of grain size, e.g., the grain size determined by conventional methods.

TABLE 10

INTERGRAIN RELAXATION CONSTANTS FOR CUBIC METALS

Metal	q	R
β-brass	2.6	0.18
Na	2.5	.16
K	2.44	.15
Pb	2.2	.10
α-brass (70–30)	2.1	.091
Cu	1.95	.071
Au	1.83	.067
Ag	1.80	.063
Fe	1.60	.040
Al	0.51	.0024
W	0.00	0.0000

APPENDIX TO SECTION A, SUBSECTION III

In this appendix the computation is given for the dissipation of energy attending the diffusion across a boundary, the diffusion distance being so small that the diffusion currents may be regarded as everywhere normal to the boundary.

The medium will be regarded as bounded on one side by the plane $x = 0$ and extending indefinitely along the positive axis. The energy, ΔE, dissipated per cycle per unit surface is then given by

$$\Delta E = \frac{2\pi}{\omega} \int_0^\infty \overline{\sigma \dot{e}}\, dx\,. \tag{i}$$

The stress and strain are related by the equation

$$de = M_R^{-1} d\sigma + \lambda dA\,, \tag{ii}$$

where the potential A is given by the diffusion equation

$$\frac{\partial A}{\partial t} = D \frac{\partial^2 A}{\partial x^2} - \gamma \dot{e} \tag{iii}$$

and by the boundary condition

$$A = 0 \quad \text{at} \quad x = 0\,. \tag{iv}$$

Corresponding to the condition of periodic vibration, we shall set

$$e \doteq e_0 \sin \omega t \,. \tag{v}$$

Then, by combining equations (i) and (ii), we obtain

$$\Delta E = -\frac{2\pi}{\omega} M_R \lambda \int_0^\infty \overline{A\,\dot{e}}\, dx \,. \tag{vi}$$

In order to evaluate this integral, we must now solve equation (iii). Toward this end we introduce the auxiliary function, $U(x, t)$, which satisfies the equation

$$\frac{\partial U}{\partial t} = D \frac{\partial^2 U}{\partial x^2}$$

and the boundary conditions

$$U(0, t) = 0 \,,$$

$$U(x, 0) = 1 \,, \; x > 0 \,.$$

The solution of equation (iii) may then be written as

$$A = \int_{-\infty}^{t} U(x, t - t') \{-\gamma \dot{e}(t')\}\, dt' \,.$$

Upon transforming

$$\tau = t - t' \,,$$

this equation becomes

$$A = -\gamma \int_0^\infty U(x, \tau)\, \dot{e}(t - \tau)\, d\tau \,. \tag{vii}$$

When we now substitute equation (v) in equation (vii), multiply by $\dot{e}$, and take the time average, we obtain

$$\overline{A\,\dot{e}} = -\frac{\gamma \omega^2 e_0^2}{2} \int_0^\infty U(x, \tau) \cos \omega\tau \, d\tau \,.$$

Upon substituting this equation into equation (vi), we obtain

$$\Delta E = -\pi \gamma \omega e_0^2 M_R \lambda \int_0^\infty \int_0^\infty U(x, \tau) \cos \omega\tau \, d\tau \, dx \,. \tag{viii}$$

It is now necessary to substitute in this equation an explicit expression for $U(x, \tau)$. This function is given in all standard treatises on heat conduction.[12] It is

$$U(x, \tau) = 2\pi^{-1/2} \int_0^{x/2\sqrt{D\tau}} e^{-\beta^2} d\beta \,. \tag{ix}$$

We first integrate with respect to τ. After integrating by parts, we obtain

$$\int_0^\infty U(x, \tau) \cos \omega\tau \, d\tau = \frac{x\omega^{-1}}{2\sqrt{\pi D}} \int_0^\infty \tau^{-3/2} e^{-x^2/4D\tau} \sin \omega\tau \, d\tau \,.$$

12. W. E. Byerly, *An Elementary Treatise on Fourier Series* (New York: Ginn & Co., 1893), Ex. 5, p. 83.

We next integrate with respect to x, obtaining

$$\int_0^\infty \int_0^\infty U(x, \tau) \cos \omega\tau \, d\tau \, dx = \omega^{-1}\left(\frac{D}{\pi}\right)^{1/2} \int_0^\infty \tau^{-1/2} \sin \omega\tau \, d\tau$$
$$= \left(\frac{D}{2}\right)^{1/2} \omega^{-3/2};$$

and therefore

$$\Delta E = \pi\lambda\gamma\left(\frac{D}{2\omega}\right)^{1/2} M_R e_0^2 .$$

If we now denote by Ω the total surface area and by V the total volume, the internal friction arising from this diffusion is

$$\tan \delta = \frac{\Delta E \cdot \Omega}{2\pi} \left(\tfrac{1}{2} M_R e_0^2 \cdot V\right)^{-1}$$

or

$$\tan \delta = \Delta_E \left(\frac{D}{2\omega}\right)^{1/2} \frac{\Omega}{V} . \qquad (x)$$

IV. RELAXATION BY THERMAL DIFFUSION

Thermal conduction provides the best-known mechanism of relaxation in metals. A rise in temperature at constant pressure is always accompanied by an increase in volume. Conversely, the adiabatic application of a tensile stress gives rise to a reduction in temperature and therefore tends to cause heat to flow into the specimen from its surroundings. As the temperature change is gradually relaxed, the specimen suffers a further slight increase in length.

The thermal relaxation strength $(E_S - E_T)/E_T$ is given, according to equation (105), by

$$\Delta_E = E_U T \frac{\alpha^2}{C_v}, \qquad (157)$$

where α is the linear thermal expansion coefficient and C_v is the specific heat per unit volume. The numerical value of Δ_E for the common metals is given in Table 11. The time of relaxation for the establishment of temperature equilibrium is given by

$$\tau \simeq \frac{d^2}{D}, \qquad (157a)$$

where d is comparable to the distance that heat must flow for the establishment of temperature equilibrium and D is the thermal diffusion coefficient given by

$$D = \frac{\sigma}{C_v}, \qquad (158)$$

where σ is the thermal conductivity, C_v, as above, is the specific heat per unit volume. The room-temperature values of D for the common metals are listed in Table 11.

The importance of thermal conduction in damping sound waves was recognized by Kirchhoff[13] as early as 1868. In all types of longitudinal vibrations, including sound, the diffusion distance is comparable with the wave length, λ. Upon utilizing the relation between λ, the frequency ν, and the velocity of sound C,

$$C = \nu\lambda ,$$

TABLE 11

THERMAL RELAXATION CONSTANTS

Metal	D (Sq. Cm/Sec)	Δ_E
Cd	0.46	0.010
Zn	0.41	.0088
Mg	0.60	.0050
Al	0.88	.0046
Be	0.53	.0046
Sn	0.40	.0040
Brass (70–30)	0.38	.0036
Ag	1.74	.0034
Cu	1.2	.0030
Ni	0.15	.0029
Pb	0.24	.0025
Fe	0.20	.0024
Pd	0.26	.0020
Sb	0.12	.0018
Au	1.1	.0017
Pt	0.26	.0015
Bi	0.065	.0014
W	0.61	.00078
Rh	0.29	.00069
Ta	0.22	0.00030

we find from equation (157*a*) that the maximum damping, which occurs when the period of vibration is comparable with the time of relaxation, appears for the wave lengths

$$\lambda \simeq \frac{D}{C} .$$

A precise computation gives

$$\lambda = 2\pi \frac{D}{C}$$

for the wave length at maximum damping. Upon taking for air the value 0.18 cm²/sec for D and 33,000 cm/sec for C, we obtain for the optimum wave length 3.5×10^{-5} cm. A similar computation for the common metals

13. G. Kirchhoff, "Über den Einfluss der Warmleitung in einen Gase auf die Schallbewegung," *Poggendorff's Ann.*, CXXXIV (1868), 177.

leads to even smaller values for the optimum wave length. Artificially produced longitudinal waves have, therefore, very little damping arising from the flow of heat along the temperature gradients produced by the waves. The waves are essentially adiabatic.

Many cases arise, however, in which temperature fluctuations are present during vibration across distances which are very short compared with the wave length of a longitudinal wave of the same frequency. The first example studied arose in transverse vibrations. Here temperature relaxation occurs through the flow of heat across the specimen. The appropriate formulae for this example are given on page 84. Here the relaxation has essentially a single time of relaxation, so a plot of internal friction versus log frequency should lead to a symmetrical bell-shaped curve, as in Figure 21. The first recorded observations on the internal friction during transverse vibration, by Bennewitz and Rötger,[14] are reproduced in Figure 25, p. 55. The vertical lines in this figure correspond to the positions of the maxima computed according to the theoretical formula,

$$\nu_0 = 2.16 \frac{D}{(\text{diameter})^2}.$$

The agreement of the computed with the observed maxima is surprisingly good, particularly in view of the fact that Bennewitz and Rötger were unaware of this formula. After learning of the theoretical interpretation[15] of their original experiments, they performed very careful experiments[16] upon one metal, German silver, in order to check the theory in all details, including the height of the maximum and the shape of the internal friction versus log frequency curve. Their results are presented in Figure 26. Nearly perfect agreement between theory and experiment was obtained.

The success of the theory of damping of transverse vibrations by transverse thermal currents led to the development of the theory[17] of damping by intercrystalline thermal currents and then to its experimental verification.[18] The theory has been reviewed on pages 84–87. The transition from the case of isothermal to the case of adiabatic vibration, with respect to

14. K. Bennewitz and H. Rötger, "Über die innere Reibung fester Körper; Absorptionsfrequenzen von Metallen in akustischen Gebiet," *Phys. Zeitschr.*, XXXVII (1936), 578.

15. C. Zener, W. Otis, and R. Nuckolls, "Experimental Demonstration of Thermoelastic Internal Friction," *Phys. Rev.*, LIII (1937), 100.

16. K. Bennewitz and H. Rötger, "Thermische Dämpfung bei Biegeschwingungen," *Zeitschr. f. tech. Phys.*, XIX (1938), 521.

17. C. Zener, "General Theory of Thermoelastic Internal Friction," *Phys. Rev.*, LIII (1938), 90.

18. R. H. Randall, F. C. Rose, and C. Zener, "Intercrystalline Thermal Currents as a Source of Internal Friction," *Phys. Rev.*, LVI (1939), 343.

adjacent grains, was achieved by varying the grain size as well as the frequency of vibration. Thus the degree of adiabaticity can depend only upon the following quantities: frequency (ν), grain size (G.S.), and thermal diffusion coefficient (D); and the only dimensionless combination of these variables is $\nu(\text{G.S.})^2/D$. As shown in Figure 27, when the observations at

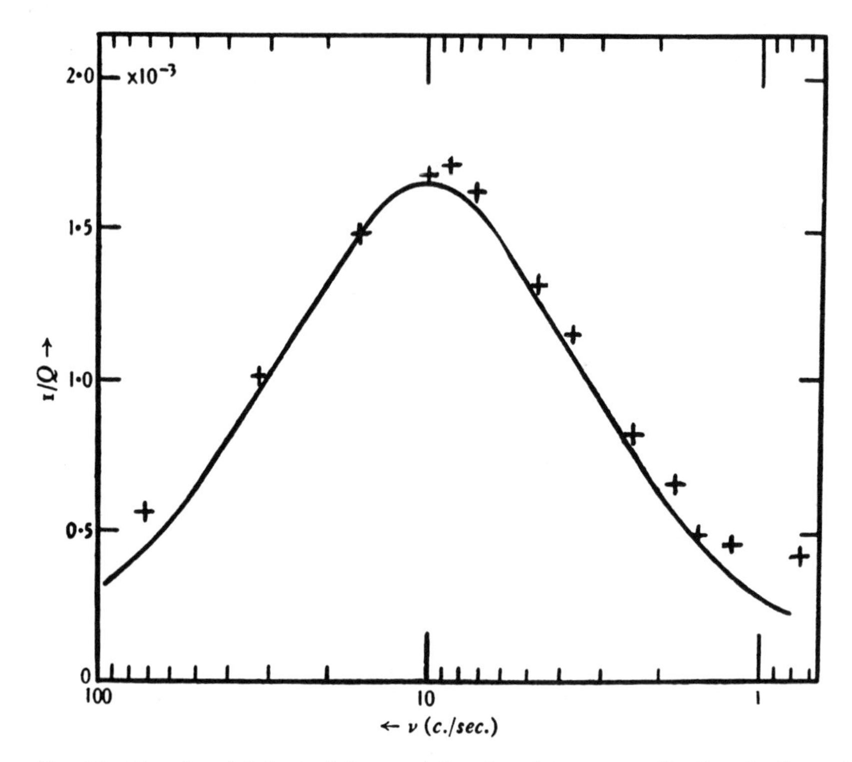

FIG. 26.—Experimental check of theory of damping of transverse vibrations by thermal currents. Example of German silver, after Bennewitz and Rötger. Curve is given by theory with no arbitrary parameters.

all frequencies and for all grain sizes are plotted against this parameter, the observations all lie upon a common curve. The maximum in the internal friction comes precisely at the theoretical[18a] value for $\nu d^2/a$ of $(3/2)\pi$. These experiments have been confirmed by Entwistle.[18b]

The maximum of the internal friction due to intercrystalline thermal currents will be greater, the greater the elastic anisotropy of the individual crystallites. The effect of this elastic anisotropy enters through the factor R in Table 10. As a check upon the theoretical dependence of the magnitude of the internal friction upon the anisotropy of the crystallites, a com-

18a. C. Zener, *Proc. Roy. Soc. London*, LII (1940), 152.

18b. K. M. Entwistle, "The Effect of Grain Size on the Damping Capacity of Alpha Brass," *Jour. Inst. Metals*, LXXV (1948), 97.

parison has been made of polycrystalline brass and aluminum. The internal friction of the latter was found[19] to be of an order of magnitude less than that of the former, in agreement with Table 10.

Finally, experiments have been performed[20] to check the theoretical predictions concerning the precise manner in which the internal friction arising from intercrystalline thermal currents varies with frequency of vibration and with grain size. It was demonstrated in section A-III of this chap-

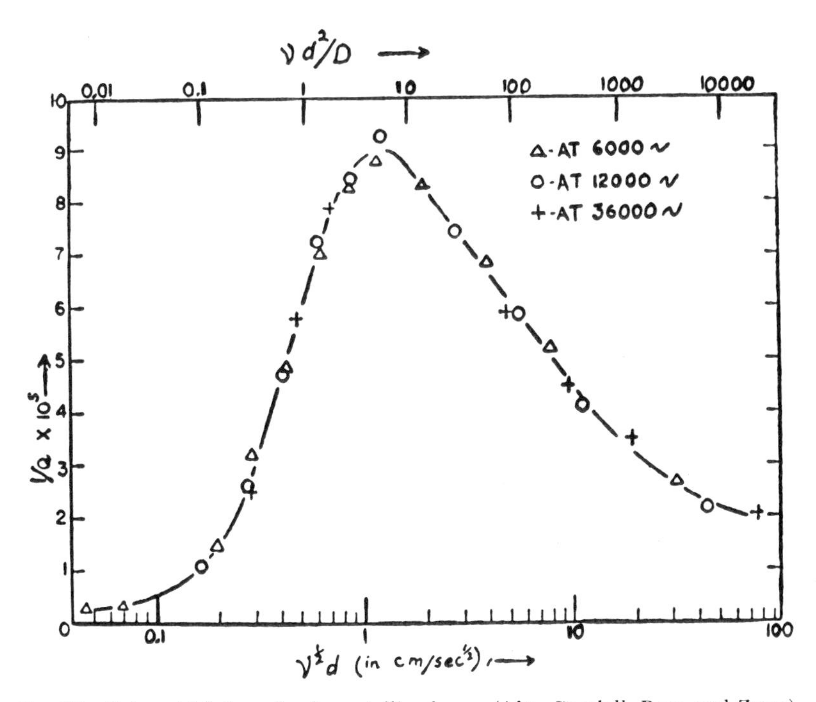

FIG. 27.—Internal friction of polycrystalline brass. (After Randall, Rose, and Zener)

ter (pp. 80–81) that in the nearly adiabatic high-frequency case the thermal currents are confined to the immediate vicinity of the grain boundaries and that in this case

$$\text{Internal friction} \sim \nu^{-1/2} .$$

The observed linearity of internal friction with $\nu^{-\frac{1}{2}}$ in the nearly adiabatic case of large grain size and high frequency is demonstrated in Figure 28. On the other hand, in the opposite extreme case of nearly isothermal vibration, with small grain size and low frequency, the total amount of heat

19. R. H. Randall and C. Zener, "Internal Friction of Aluminum," *Phys. Rev.*, LVIII (1940), 472.

20. C. Zener and R. H. Randall, "Variations of Internal Friction with Grain Size," *Trans. A.I.M.E.*, CXXXVII (1940), 41.

which flows between adjacent grains per half-cycle is nearly independent of frequency. The thermal current is therefore proportional to the frequency, and the rate of degradation of energy by thermal flow is proportional to the square of the frequency. The mechanical energy dissipated per cycle and hence the internal friction are therefore in this extreme case proportional to the first power of the frequency. The experimental demonstration of this deduction is given as Figure 29.

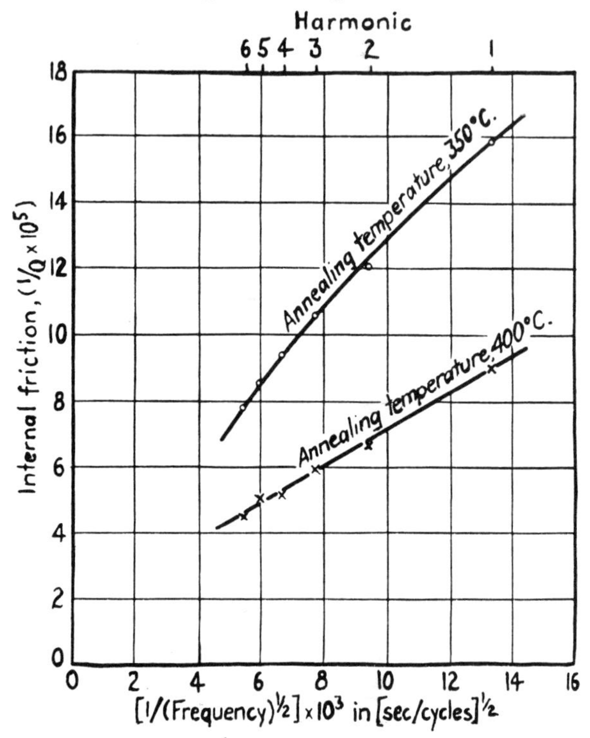

FIG. 28.—Example of (frequency)$^{-\frac{1}{2}}$ dependence of internal friction in extreme adiabatic case. (After Zener and Randall.)

We shall now return to the subject of the damping of longitudinal waves by the flow of heat from regions of compression to regions of dilation. It was pointed out that the optimum wave length, λ_0, for maximum damping was extremely short and that therefore such damping is of no importance in artificially produced waves. The possibility arises, however, that this damping may be of importance to the Debye heat waves. This possibility is examined below.

Shortly after introducing his now well-known theory of specific heat, Debye[21] pointed out that the thermal waves in nonconductors must be

21. M. Planck, P. Debye, *et al.*, *Vorträge über die kinetische Theorie der Materie und der Electrizität* (Leipzig: Teubner, 1914).

coupled with each other in order that the thermal conductivity may be finite. If there were no such coupling, the thermal conductivity would be infinite. The coupling resides, of course, in the variation of frequency with dilation. No unique method exists, however, for using this coupling to find out how the waves interact. Debye introduced the viewpoint that the thermal waves give rise to fluctuations in density and that it is these density fluctuations which scatter the heat waves in a manner analogous to

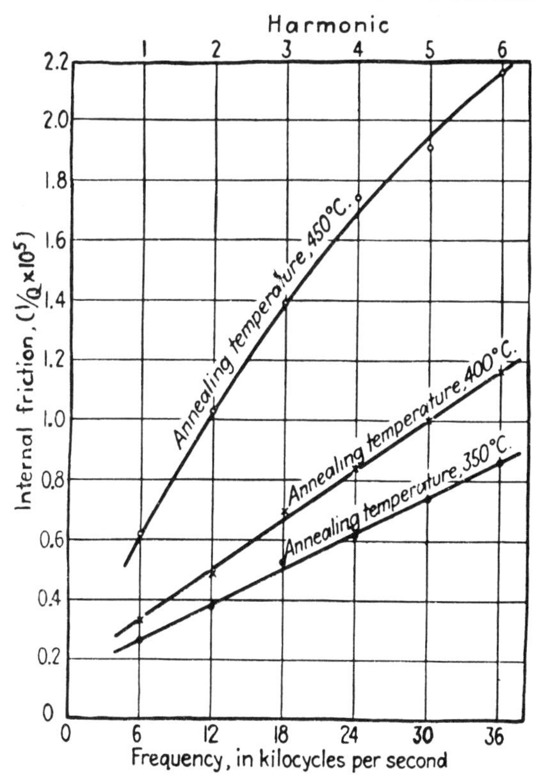

FIG. 29.—Example of linear dependence of internal friction upon frequency in extreme isothermal case. (After Zener and Randall.)

that by which fluctuations in the density of the air scatter light waves. R. Peierls[22] has pointed out that this viewpoint leads to an intrinsic difficulty, namely, the scattering cross-section of the density fluctuations decreases with the frequency of the waves as ν^4, and therefore the mean free path would be proportional to ν^{-4}. Since the frequency density of the waves varies as only ν^2, the integrand for the mean free path will involve ν^{-2}. The Debye viewpoint would therefore lead to an infinite thermal conductivity arising from the low-frequency thermal waves.

22. "Zur kinetische Theorie der Warmleitung in Kristalle," *Ann. d. Phys.*, III (1929), 1055.

The above-mentioned difficulty in the Debye viewpoint disappears as soon as we recognize that the low-frequency, long-wave-length thermal waves may be damped by the heat currents which they generate. For frequencies much less than the optimum frequency, ν_0, the internal friction due to the thermoelastic coupling varies as ν. The mean free path of these waves will therefore vary as ν^{-2}. The integrand for the average mean free path in the long-wave-length region, therefore, does not diverge.

V. RELAXATION BY ATOMIC DIFFUSION

In an unstressed solid solution, equilibrium is attained only when the solute atoms are randomly distributed. When stress fluctuations are present within a system, equilibrium conditions no longer correspond to a uniform distribution of solute atoms. Thus, if an increase in concentration is attended by a lattice expansion, then under equilibrium conditions a dilation will be attended by an increase in concentration. Fluctuations in dilation will therefore tend to induce fluctuations in concentration. In other words, fluctuations in dilation will induce fluctuations in the partial thermodynamic potential, G_k, of the solute k, fluctuations which are relaxed only by atomic diffusion. That such diffusion will lead to anelastic effects was first recognized by Gorsky,[23] who formulated the complete theory for the effects of this diffusion. No experimental study has as yet been made on anelastic effects resulting from atomic diffusion of either a macroscopic or a microscopic type.

A characteristic feature of relaxation by atomic diffusion is the relatively long time of relaxation and the rapid dependence of this time of relaxation upon temperature. The relaxation time, τ, will be given approximately by

$$\tau \simeq \frac{d^2}{D}, \tag{159}$$

where d is the mean distance over which diffusion must take place for relaxation and D is the atomic diffusion coefficient. This diffusion coefficient may be written as

$$D \simeq \frac{a^2}{\tau_0}, \tag{160}$$

where a is the distance that an atom moves in an elementary diffusion act and τ_0 is the average time interval between elementary diffusion acts. Combining the above two equations, we obtain

$$\tau \simeq \left(\frac{d}{a}\right)^2 \tau_0. \tag{161}$$

23. W. S. Gorsky, "On the Transitions in the CuAu Alloy. III," *Phys. Zeitschr. Sow.*, VI (1936), 77.

The rapid temperature variation of the relaxation time, τ, comes about through the rapid temperature variation of τ_0. According to all theories of diffusion, the mean time interval, τ_0, depends upon temperature essentially as exp (H/RT), where the quantity H is known as the "heat of activation" and R is the gas constant. According to the well-known Langmuir-Dushman[24] theory, the coefficient is given by approximately (hN/H) where h is Planck's constant and N is Avogadro's number. Thus

$$\tau_0 \simeq \frac{hN}{H} e^{H/RT}. \tag{162}$$

The known values of the heats of activation of elements in metals are given in Table 12. As an example of the exceedingly long times required for an elementary diffusion act at room temperature, the following examples will be quoted for the room temperature values of τ_0 as derived from equation (162): 1.6, 2×10^7, 0.5×10^{15} seconds for H = 20,000, 30,000, 40,000 cal/mole, respectively.

The relative change in the elastic modulus associated with relaxation of all changes in the partial thermodynamic, G_k, is, according to equation (111a),

$$\Delta_E = E_U \frac{\left(\frac{\partial e}{\partial G_k}\right)^2}{\left(\frac{\partial n_k}{\partial G_k}\right)}. \tag{163}$$

Here n_k is the number of solute atoms of type k per unit volume. When the solute atoms form an interstitial solution, the partial thermodynamic potential will be defined in the standard manner, namely, as

$$G_k = \frac{\partial G}{\partial n_k}, \tag{164}$$

where G is the free energy per unit volume. A different definition will be adopted in the case of solutes which form substitutional lattices, since in such cases diffusion of the solute atoms necessitates a counterdiffusion of the solvent atoms. Here the partial derivative in equation (164) implies the change in G induced by the addition of one solute atom with the simultaneous subtraction of a solvent atom. After using the transformation,

$$\frac{\partial e}{\partial G_k} = \frac{\partial e}{\partial n_k} \cdot \frac{\partial n_k}{\partial G_k},$$

24. S. Dushman and I. Langmuir, *Phys. Rev.*, XX (1922), 113.

TABLE 12

HEATS OF ACTIVATION OF ATOMIC DIFFUSION COEFFICIENTS*

Solvent	Solute	H (Cal/Mole)
Ag	Au	29,800
	Cd	22,350
	Cu	24,800
	Sb	21,500
	Sn	21,400
Al	Cu	34,900
	Mg	38,500
Au	Au	51,000
	Cu	27,400
	Pd	37,400
Cu	Al	40,200
	Be	27,900
	Cu	57,200
	Si	39,950
	Zn	42,000
Fe (α)	C†	18,000
	N‡	20,000
	Fe§	73,000
Fe (γ)	C‖	32,000
	N	34,000
	Fe§	74,000
Ta¶	C	25,000
	O	29,000
Pb	Au	13,000
	Bi	18,600
	Hg	19,000
	Pb	27,900
	Sn	24,000
W	Mo	80,500
	Th	120,000
	U	100,000

* Unless otherwise stated, data are from R. M. Barrer, *Diffusion in and through Solids* (New York: Macmillan Co., 1941).

† J. L. Snoek, *Physica*, VIII (1941), 711.

‡ T. S. Kê, *Metals Tech.* (1943). (In Press.)

§ C. Birchenall and R. Mehl, *J. App. Phys.*, XIX (1948), 217.

‖ C. Wells and R. Mehl, *Trans. A.I.M.E.*, CXL (1940), 294.

¶ T. S. Kê, *Phys. Rev.* (1948). (In Press.)

and equation (164), we may re-write equation (163) as

$$\Delta_E = E_U \frac{\left(\frac{\partial e}{\partial n_k}\right)^2}{\left(\frac{\partial^2 G}{\partial n_k^2}\right)}. \tag{165}$$

In the case of a dilute solution or, more precisely, when the total free energy minus the free energy of mixing is a linear function of the number of the various types of atoms, the denominator of equation (165) may be further reduced. In this case

$$G_k = kT \{ ln\ n_k - ln\ (N - n_k) \} , \tag{166}$$

where N is the total number of atoms per unit volume in the case of a substitutional solute, the total number of interstitial positions in the case of interstitial solutions. Upon defining the atomic concentration, X, as

$$X = \frac{n_k}{N},$$

we obtain

$$\frac{\partial^2 G}{\partial n^2} = \frac{kT}{N} X^{-1} (1 - X)^{-1} \tag{167}$$

and therefore

$$\Delta_E = \frac{X(1 - X)}{kT} \cdot \frac{E_U}{N} \left(\frac{\partial e}{\partial X}\right)_\sigma^2. \tag{168}$$

The second factor is of the order of magnitude of the strain energy introduced by a single solute atom.

For purposes of computation it is convenient to replace N in equation (168) by $(\rho N_0/W)$, where ρ is the density, N_0 is Avogadro's number, and W is the atomic weight. Then, upon replacing $N_0 k$ by R, the gas constant, we obtain

$$\Delta_E = X(1 - X) \frac{W E_U}{\rho R T} \left(\frac{\partial e}{\partial X}\right)^2. \tag{169}$$

We shall now estimate the maximum value of Δ_E that might be expected. The product $X(1-X)$ has the maximum value of $\frac{1}{4}$. For the third factor we shall take copper as a typical solvent and shall consider 700° C. as a typical temperature, obtaining 210 for this factor. In estimating the maximum value of the last factor, we combine the approximate relation

$$\frac{\partial e}{\partial X} \simeq \frac{a_2 - a_1}{a_1},$$

where a_2 and a_1 are the atomic radii of the solute and solvent atoms, respectively, with Hume-Rothery's rule,[25, 26]

$$\frac{a_2 - a_1}{a_1} < 0.14 \, ,$$

and obtain 0.02 as the maximum value of the last factor. Combining the above approximations, we have

$$\Delta_E < 1 \, . \qquad (170)$$

The low atomic diffusion rates render it unlikely that examples will ever be observed in which the modulus is appreciably relaxed through macroscopic atomic diffusion. The relaxation occasioned by diffusion between adjacent grains is smaller than that for macroscopic diffusion, by the ratio R given in Table 10.

In many cases the assumption of a dilute solution cannot be used. In such cases the relaxation strength is more conveniently written as

$$\Delta_E = E_U \frac{\left(\frac{\partial e}{\partial X}\right)_\sigma^2}{\left(\frac{\partial^2 G}{\partial X^2}\right)_\sigma} \, . \qquad (171)$$

Two extreme cases discussed below deserve special study. In one the curvature $\partial^2 G/\partial X^2$ is much smaller than that given by the assumption of a dilute solution; in the other it is much larger.

Certain pairs of metals are completely soluble in each other at elevated temperatures, while a limited region of immiscibility exists below a certain critical temperature. An example is given in Figure 30. Now the existence of a single phase throughout the entire concentration range implies that the curvature of $G(X)$ is everywhere positive—in other words, that a given line can be tangent to $G(X)$ at only one point. Such is the case in the lower curve of Figure 31. On the other hand, the existence of a two-phase region implies that a line can be drawn which is tangent to $G(X)$ at two points, as illustrated by the upper curve of Figure 31. It is readily seen that immiscibility implies that, over a limited concentration range, $\partial^2 G/\partial X^2$ is negative. At the critical temperature it is therefore necessary that $\partial^2 G/\partial X^2$ be zero at a certain concentration. For a critical concentration the relaxation strength associated with atomic diffusion therefore approaches infinity as the temperature is lowered to the critical temperature. Mayer[27] has, in fact, argued that $\partial^2 G/\partial X^2$ will be zero over a finite range of concentration and of temperature.

25 W. Hume-Rothery, G. W. Mabbot, and K. M. Channel-Evans, *Phil. Trans. Roy. Soc., A*, CCXXXIII (1934), 44.

26. W. Hume-Rothery, *The Structure of Metals and Alloys* (London, 1936), p. 59.

27. J. Mayer and M. Mayer, *Statistical Mechanics* (New York: John Wiley & Sons, Inc., 1940), pp. 312–13.

In the second extreme case the curvature $\partial^2 G/\partial X^2$ is much larger than would be given by the assumption of a dilute solution. This curvature is always anomalously large in those phases which have a narrow-solution range. As may be seen from Figure 32, a narrow-solution range at all tem-

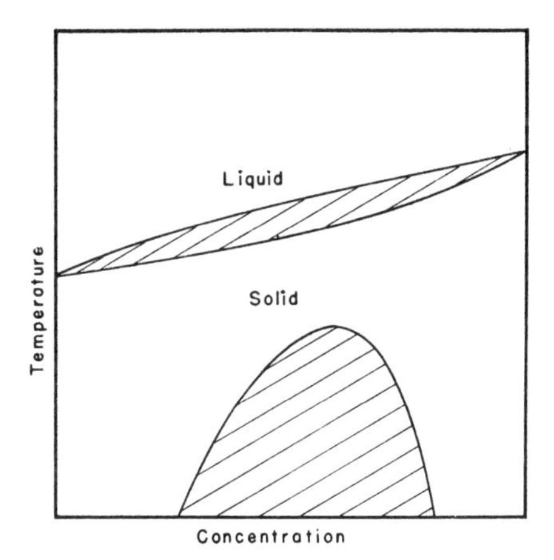

FIG. 30.—Example of a critical temperature for immiscibility

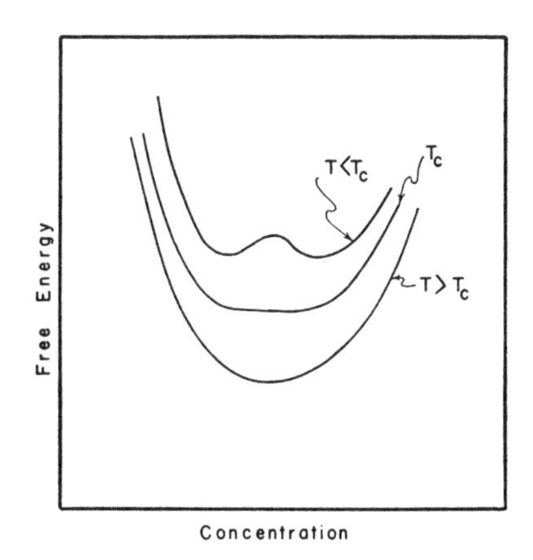

FIG. 31.—Variation of free energy near critical temperature

peratures necessarily implies a large curvature. Relaxation effects associated with atomic diffusion are therefore anomalously small in such phases.

VI. RELAXATION BY MAGNETIC DIFFUSION

It has been recognized for over a century that a coupling exists between the magnetic and the mechanical properties of metals. Joule[28] reported in

28. J. P. Joule, "On the Effects of Magnetism upon the Dimensions of Iron and Steel Bars," *Phil. Mag.*, Vol. LXXVI (1847).

1847 that the magnetization of iron and steel resulted in changes in dimensions; and in the same year Matteuci[29] reported that changes in dimensions resulted in a change in magnetization. Through this coupling all types of magnetic relaxation give rise to anelasticity.

The type of magnetic relaxation most amenable to theoretical treatment is that which results from the macroscopic electrical eddy currents, which tend to shield the interior of a ferromagnetic material from changes in magnetic flux. Thus, suppose a tensile stress is suddenly applied to a very long and thin magnetized specimen, thereby suddenly changing the intensity of magnetization. This change in magnetization induces eddy

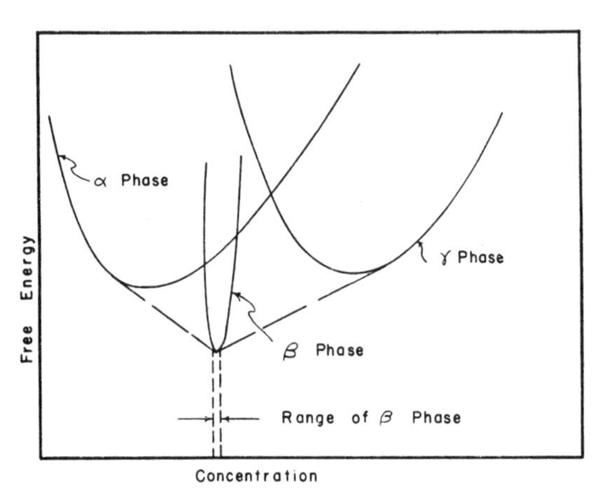

FIG. 32.—Illustration of how a large curvature $\partial^2 G/\partial X^2$ is related to a narrow range of solubility.

currents at the surface of the specimen of such a magnitude that the magnetic flux within the specimen is unchanged. In other words, the magnetic-field intensity induced by the eddy currents is of just such a magnitude that the change in magnetic flux density,

$$\delta B = \delta H + 4\pi\, \delta M\ ,$$

is exactly zero. The eddy currents gradually diffuse inward from the surface, thereby allowing magnetic flux to diffuse inward, at the same time reducing the magnetic-field strength to its original value. In such a case the magnetic-field strength, H, may be regarded as a thermodynamic potential, the magnetic flux density as a thermodynamic density in the sense of Section I of this chapter.

The justification of the above viewpoint may be obtained directly from

29. Matteuci, *Compt. rend. Acad. Sci.*, 1847.

Maxwell's equations. Neglecting the polarization current density compared with the true current density within the metal, one obtains

$$\frac{d\boldsymbol{B}}{dt}=\frac{10^8}{0.4\pi \mathrm{K}}\,\mathrm{curl}\;\;\mathrm{curl}\,\boldsymbol{H}\,, \tag{172}$$

where K is the electrical conductivity in ohm^{-1} cm.$^{-1}$ When the stress, and therefore $\boldsymbol{B}$ and $\boldsymbol{H}$, suffer small periodic fluctuations, we may set

$$dB_x=(C_{11}d\sigma_{xx}+C_{12}d\sigma_{xy}+\ldots)+\mu dH_x\,,$$

with analogous equations for dB_y and dB_z. If the stress components are uniform over the specimen, as in tension, or vary, at most, in a linear manner, as in the torsion of a cylindrical rod or in bending, the substitution of these equations in equation (172) and the use of the condition that the divergence of $\boldsymbol{B}$ is zero lead to

$$\frac{d\boldsymbol{B}}{dt}=D\nabla^2\boldsymbol{B}\,, \tag{173}$$

where the magnetic diffusion coefficient, D, is given by

$$D=\frac{10^8}{0.4\pi \mathrm{K}\mu}\,. \tag{174}$$

According to this equation, instantaneous changes in stress leave the magnetic flux density unchanged, the latter changing only by the diffusion of flux from the outside, the diffusion coefficient being given by equation (174).

The change in energy per unit volume associated with a change in magnetic flux density is $\boldsymbol{H}\cdot(d\boldsymbol{B}/4\pi)$. Therefore, in the case of a long, thin specimen, when $\boldsymbol{H}$ and $\boldsymbol{B}$ are parallel to the axis of the specimen, if H is regarded as a thermodynamic potential, $B/4\pi$ must be regarded as the corresponding density. Hence the equations relating strain, e, and magnetic flux density, B, to tensile stress, σ, and magnetic-field strength, H, are

$$de=E_R^{-1}d\sigma+\lambda dH\,,$$

$$dB=4\pi\lambda d\sigma+\mu dH\,.$$

The strength of the relaxation due to magnetic flux diffusion, $(E_B-E_H)/E_H$, is computed from these equations as

$$\Delta_E=4\pi\mu^{-1}\lambda^2E_U\,. \tag{175}$$

The magnitude of the relaxation strength, Δ_E, is a function of the state of magnetization. It is zero at zero magnetization, for at zero magnetization a stress can induce no change in magnetization. The relaxation strength is also zero at saturation magnetization; for, at saturation, stress can likewise produce no change in magnetization. The maximum value of Δ_E will be in the vicinity of remanence magnetization. Becker and Döring[30] have shown that, at remanence,

$$\Delta_E \simeq 0.3\lambda_s \frac{E_U}{\sigma_i}, \tag{176}$$

where λ_s is the tensile strain which accompanies a change in magnetization from zero to saturation, and σ_i is a measure of the residual microstresses, which may have arisen either through prior plastic deformation or through the presence of impurity atoms. In pure well-annealed ferromagnetic materials the only residual stresses are those arising from magnetostriction. In this case

$$\sigma_i \simeq \lambda E_U ;$$

and therefore

$$\Delta_E \simeq 0.3 . \tag{177}$$

No observations have been made on the magnitude of Δ_E associated with the macroscopic diffusion of magnetic flux. Such observations would involve a comparison of the elastic moduli measured under conditions of essentially constant B, such as at high frequencies or upon thick specimens, with the values measured under conditions of essentially constant H, such as at low frequency or upon thin specimens.

The effective time of relaxation, τ, for magnetic diffusion is given approximately by the equation

$$\tau \simeq \frac{d^2}{D}, \tag{178}$$

where d is a transverse dimension of the specimen, and the magnetic diffusion coefficient, D, is given by equation (174). In this equation, μ refers to the reversible permeability, which ranges from 20 for the common steels to 14,000 for carefully prepared high-purity iron. Over this range of reversible permeability, D changes from 40 to 0.05 cm^2/sec. In the high-frequency dynamical methods commonly used to determine the elastic moduli of steels, the measurements are therefore made essentially under conditions of constant B. The internal friction arising from the macroscopic diffusion of B is at a maximum when the period of vibration is comparable to τ, as given by equation (178). The precise manner in which the internal friction varies with frequency was first computed by Kersten[31]

30. R. Becker and W. Döring, *Ferromagnetismus* (Berlin, 1939; New York: Edwards Bros., 1943), p. 378.

31. M. Kersten, "Zur Deutung der mechanischen Dämpfung ferromagnetischen Werkestoffe bei Magnetisierung," *Tech. Phys.* XV (1934), 463.

for the case of longitudinal vibration in specimens of circular cross-section and by the author[32] for other examples. The formulae have been given in Section III*a* (pp. 81-83).

Honda[33] discovered what he called, and what is now commonly known as, the "ΔE-effect." The term "ΔE" refers to the difference in the elastic modulus of a specimen in the demagnetized state and at magnetic saturation. In the demagnetized state the elementary magnetic domains are free to change their direction of magnetization under the action of an applied stress and to grow at the expense of neighboring domains having a different orientation. At magnetic saturation such changes are no longer possible. A complete theory of the ΔE-effect was developed by Akulov and Kondorsky.[34] Their formula is

$$\frac{\Delta E}{E} = \tfrac{3}{5}\,\chi_0 \lambda_{100}^2\,\frac{E}{J_s^2}\,.$$

Here χ_0 is the magnetic susceptibility at zero magnetization, λ_{100} is the longitudinal magnetostriction for a single crystal saturated in the [100] direction, and J_s is the saturation magnetization. In their derivation of the above equation, Akulov and Kondorsky assumed constant stress throughout the specimen. Cooke and Brown[35] have subjected this formula to a careful experimental test. They find that their observed value of $\Delta E/E$ lies between that given by the above formula, in which constant stress was assumed, and a formula obtained under the assumption of constant strain throughout the specimen. A review of the many experiments upon the ΔE-effect has been given by Becker and Döring.[36] They have shown that in soft magnetic specimens, where the only residual stresses are due to magnetostriction, $\Delta E/E$ is of the order of magnitude of 0.3. Values as high as 0.17 have been observed in nickel.[37]

As previously mentioned, relaxation by macroscopic diffusion of magnetic flux does not occur either in the demagnetized state or at magnetic saturation. Hence such diffusion does not contribute to the ΔE-effect. If, however, the elastic modulus is measured as a function of magnetization, from zero magnetization to saturation, the macroscopic diffusion of mag-

32. C. Zener, "General Theory of Macroscopic Eddy Currents," *Phys. Rev.*, LIII (1938), 1010.

33. K. Honda, S. Shimizer, and S. Kusakatee, "Veränderung des Elastizitätskoeffizienten ferromagnetischer Substanzen infolge von Magnetisierung," *Phys. Zeitschr.*, III (1901–2), 380, 381.

34. Akulov and Kondorsky, *Zeitschr. f. Phys.* LXXVIII (1932), 801; LXXXV (1933), 661.

35. W. T. Cooke and W. F. Brown, "Variation of the Internal Friction and Elastic Constants with Magnetization in Iron," *Phys. Rev.*, L (1936), 1158, 1165.

36. *Op. cit.*

37. M. Nakamura, *Sci. & Tech. Repts., Tohoku Univ.*, XXIV (1935), 302; *Zeitschr. f. Phys.*, XCIV (1935), 707.

netic flux will contribute to changes in the modulus at some intermediate range of magnetization.

VII. RELAXATION OF ORDERED DISTRIBUTIONS

In some crystals, such as ionic crystals, the position of each type of atom or ion is uniquely determined. Thus in CsCl, the Cs ions are located at the lattice points of one simple cubic lattice, the Cl ions are at the lattice points of a second simple cubic lattice. In other crystals, of which β-brass of the composition CuZn is an example, each type of atom manifests a preference for certain lattice positions below a certain critical temperature. Thus below 470° C. the copper atoms in CuZn have a preference for the lattice positions of one simple cubic lattice, the zinc atoms have a preference for the lattice positions of the other simple cubic lattice. Such solids are said to be "ordered" below their critical temperature. Further examples of ordered metals, as well as a review of their properties, have been given in several recent reviews.[38] Ordered structures have two properties which render them of special interest to the present discussion: (1) below the critical temperature the degree of order is incomplete, the equilibrium order becoming more complete as the temperature is lowered; and (2) a change in lattice dimensions accompanies changes in degree of ordering. As was first pointed out by Gorsky,[39] a change in lattice dimensions by an externally applied stress will therefore change the equilibrium order, and the time delay associated with the establishment of equilibrium will lead to anelastic effects.

The quantitative analysis of the anelasticity associated with ordering may most conveniently be approached by considering the entropy of a unit volume to consist of two parts. The first part pertains to the uncertainty in the positions of the individual atoms with regard to their respective equilibrium positions and will be referred to as the "lattice entropy," S_L. The second part pertains to the uncertainty in the equilibrium positions of the atoms and will be denoted by S_M. This second part is commonly referred to as the "entropy of mixing."[40] Corresponding to these two entropies, there are two temperatures, T_L and T_M, defined as follows:

$$T_L = \left(\frac{\partial H}{\partial S_L}\right)_\sigma ,$$

$$T_M = \left(\frac{\partial H}{\partial S_M}\right)_\sigma ,$$

38. Borelius, *Zeitschr. f. Electrochemie*, XLV (1939), 16; and F. C. Nix and W. Shockley, "Order-Disorder Transformations in Alloys," *Rev. Mod. Phys.*, X (1938), 1.

39. W. S. Gorsky, "Die elastische Nachwirkung in geordneter CuAu Legierung," *Phys. Zeitschr. Sow.*, VIII (1935), 562.

40. R. H. Fowler and E. A. Guggenheim, *Statistical Thermodynamics* (Cambridge: Cambridge University Press, 1939), p. 163.

where H is the heat content function per unit volume. The entropy of mixing, S_M, can increase or decrease only through a transfer of heat from the vibrational coordinates to the coordinates of mixing or vice versa. The anelasticity associated with ordering may therefore be regarded as arising from the flow of heat between the vibrational coordinates and the mixing coordinates. In computing the relaxation strength associated with ordering, due regard must therefore be given to the conditions which T_L satisfies.

Consider, as a first example, that T_L remains essentially constant, either through maintenance of thermal equilibrium with the surroundings or through maintenance of thermal equilibrium between different parts of the specimen, as in transverse flexure. Then, from equation (105),

$$\Delta_E = E_U \mu_M^{-1} \lambda_M^2 , \tag{179}$$

where

$$\mu_M = \left(\frac{\partial S_M}{\partial T_M}\right)_\sigma \tag{180}$$

and

$$\lambda_M = \left(\frac{\partial e}{\partial T_M}\right)_\sigma . \tag{181}$$

An alternative expression is obtained from equation (105a), namely,

$$\Delta_E = E_U \mu_M \left(\frac{\partial e}{\partial S_M}\right)_\sigma^2 . \tag{182}$$

Those parts of the thermal expansion and of the heat content which are associated with disordering vary with temperature in approximately the same manner. An example is given in Figures 33 and 34 for the alloy Cu_3Au, the data for the specific heat being taken from Sykes and Jones[41] and those for the thermal expansion coefficient from Nix and MacNair.[42] The parallel behavior of the anomaly in thermal expansion and in heat content suggest that the derivative $(\partial e/\partial S_M)_\sigma$ is essentially independent of temperature. We shall therefore assume

$$\left(\frac{\partial e}{\partial S_M}\right)_\sigma = \frac{\Delta e}{\Delta S_M},$$

where Δe is the strain associated with the order-disorder transformation and ΔS_M is the corresponding change in entropy. Thus

$$\Delta_E = E_U \mu_M \left(\frac{\Delta e}{\Delta S_M}\right)^2 . \tag{183}$$

41. C. Sykes and F. W. Jones, "The Atomic Rearrangement Process in the Copper-Gold Alloy Cu_3Au," *Proc. Roy. Soc., London, A*, CLVII (1936), 213.

42. F. C. Nix and D. MacNair, "A Dilatometric Study of the Order-Disorder Transformation in CuAu Alloys," *Phys. Rev.*, LX (1941), 320.

The entropy change, ΔS_M, may be readily estimated from theoretical considerations. Thus let the alloy have the composition A_rB_s. If a unit volume contains N atoms, the entropy of mixing in the disordered state, regarding all atoms of a like kind as indistinguishable, is $k\ln\{N!/(N-pN)!(N-qN)!\}$, where p and q are the ratios $r/(r+s)$ and $s/(r+s)$, respectively. In the completely ordered state the entropy of mixing is zero. One therefore obtains

$$\Delta S_M = -Nk\{p\ln p + q\ln q\}. \tag{184}$$

Thus

$$\Delta S_M = \begin{cases} 0.64Nk, (r, s) = (1, 1), \\ 0.56Nk, (r, s) = (1, 3). \end{cases} \tag{185}$$

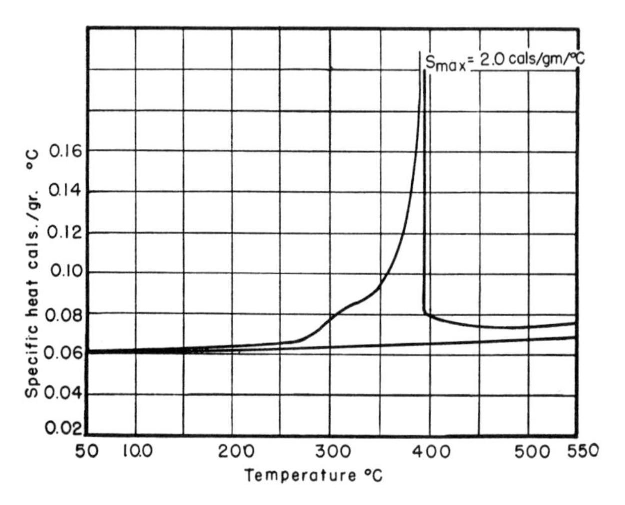

FIG. 33.—Specific heat of Cu_3Au. (After Sykes and Jones)

Equations (183) and (185) will now be applied to the case of Cu_3Au. For this case Δe is 0.0019,[43] and, from equation (185) ΔS_M is 0.56 Nk. We find

$$\Delta_E = 0.004, \text{ when } \mu_M = \mu_L .$$

Just below the critical temperature, μ_M is at least thirty times as large as μ_L. Thus just below the critical temperature Δ_E rises to a value as large as 0.1.

As previously mentioned, equation (183) is applicable to those cases in which the temperature of the lattice vibrations, T_L, remains essentially constant. These conditions are most readily obtained by flexure of thin specimens, where the fluctuations in T_L are rapidly equalized by transverse thermal conduction. It is, of course, necessary that the relaxation time for equalization of T_L across the specimen be short compared with

43. *Ibid.*

the relaxation time for the equalization of T_M and of T_L. Elastic aftereffects under these conditions have been reported by Gorsky[44] for CuAu. Gorsky did not find that this alloy behaved as if it had a single time f relaxation, and he believed a second type of relaxation to be present. Since CuAu forms a tetragonal lattice in the ordered state, with a tetragonality ratio differing only slightly from unity, it is possible that this second type of relaxation is associated with the movement of twin interfaces.

In many cases the ordered structure has a tetragonal or a rhombohedral symmetry, although the disordered state has a cubic symmetry. The alloy CuAu is an example. In these cases Δe^2 in equation (183) must be replaced by

$$(\Delta e)^2_{\text{av}} = \tfrac{1}{3}\,\Delta e^2_{11} + \tfrac{2}{3}\,\Delta e^2_{\perp},$$

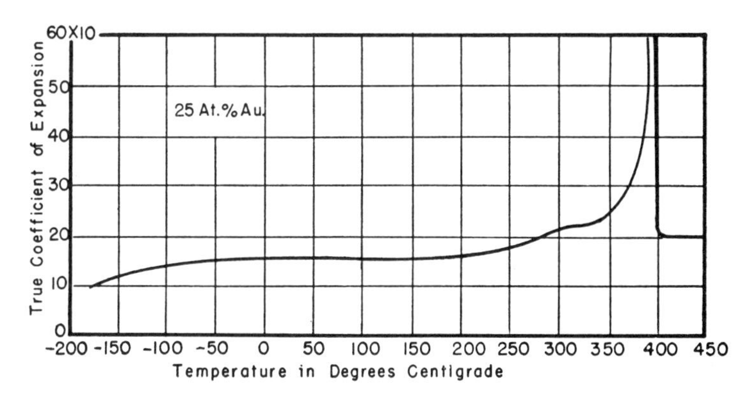

FIG. 34.—Thermal expansion coefficient of Cu_3Au. (After Nix and MacNair)

where Δe_{11} and $\Delta e_{\perp}$ are the strains introduced by ordering parallel and normal, respectively, to the principal axis. In the case of CuAu, these strains are 0.07 and 0.035, respectively.[45] The quantities $(\Delta e)^2_{\text{av}}$ and therefore Δ_E are hence two orders of magnitude greater than in the case of the cubic-ordered alloy, Cu_3Au.

The preceding discussion has been limited to the case in which the temperature associated with the lattice vibrations, T_L, remains essentially constant. In this case Δ_E refers to the complete relaxation of all fluctuations in the temperature associated with ordering, T_M. We shall now consider the case in which no heat flows in or out of an element of volume. The only exchange of heat will then be between the mixing coordinate and the lattice vibrations. In order to compute the associated relaxation

44. Gorsky, *Phys. Zeitschr. Sow.*, VIII (1935), 562.
45. *Ibid.*

strength, we start from the following equations, which relate the dependent variables e, S_M, and S_L with the independent variables σ, T_M, and T_L:

$$\left.\begin{aligned} de &= E_R^{-1} d\sigma + \lambda_M dT_M + \lambda_L dT_L\,, \\ dS_M &= \lambda_M d\sigma + \mu_M dT_M\,, \\ dS_L &= \lambda_L d\sigma + \mu_L dT_L\,. \end{aligned}\right\} \tag{186}$$

In writing these equations it has been assumed that $(\partial S_M/\partial T_L)_{\sigma,\,T_M}$ is of negligible magnitude. This assumption seems reasonable, since T_L can influence S_M only through the thermal expansion associated with a change in T_L. We are now interested in the relative difference in Young's modulus in the two cases in which S_M and S_L are both held constant and in which the sum of S_M and S_L is held constant and T_M and T_L are kept equal to one another by a flow of heat between the mixing coordinates and the lattice vibrations. In other words, we shall now compute

$$\Delta_E = \frac{\left(\dfrac{\partial\sigma}{\partial e}\right)_{S_M,\,S_L} - \left(\dfrac{\partial\sigma}{\partial e}\right)_{S_M+S_L,\;T_M=T_L}}{\left(\dfrac{\partial\sigma}{\partial e}\right)_{S_M+S_L,\;T_M=T_L}}\,.$$

A straightforward computation based upon the set of equations (186) leads to

$$\Delta_E = E_U\,\frac{\mu_M\mu_L}{\mu_M+\mu_L}\left\{\left(\frac{\partial e}{\partial S_M}\right)_{\sigma,\,T_L} - \left(\frac{\partial e}{\partial S_L}\right)_{\sigma,\,T_M}\right\}. \tag{187}$$

In order to see the effect upon Δ_E of the restraint of T_L, equation (187) should be compared with equation (182). When $\mu_M = \mu_L$, the second factors differ by a factor of 2. At least in the case of Cu_3Au, the third factor in equation (187) is two orders of magnitude less than the corresponding factor in equation (182). This result is a consequence of the close identity of the two terms in the third factor of equation (187). In order to interpret this result, we observe that

$$\left(\frac{\partial e}{\partial S_M}\right)_{\sigma,\,T_L} = \left(\frac{\partial T_M}{\partial\sigma}\right)_{S_M,\,T_L} = \left(\frac{\partial T_M}{\partial\sigma}\right)_{S_M,\,S_L}, \tag{188}$$

which follows directly from equation (186). The close identity of the two terms in the last factor of equation (187) therefore implies that when a tensile stress is applied to the metal under strictly adiabatic conditions with respect to all types of heat transfer, T_M and T_L are increased by essentially the same amount.

Observations have been made by Köster[46] upon the internal friction of

46. W. Köster, *Zeitschr. f. Elektrochemie*, XLV (1939), 31.

β-brass under conditions where $S_M + S_L$ was essentially constant. His observations show a maximum near the critical temperature. One must surmise from these results that the two derivatives $\partial e/\partial S_M$ and $\partial e/\partial S_L$ in equation (187) in no wise cancel each other in β-brass as they do in Cu_3Au.

The only study which has been made of the time of relaxation for the establishment of equilibrium order is that of Gorsky[47] upon CuAu. He finds this time of relaxation to obey the relation

$$\tau \sim e^{Q/RT} ,$$

where the heat of activation, Q, has the same value as for macroscopic diffusion, namely, 27,400 cal/mole. At 275° C., which is about 100° C. below the critical temperature, this time of relaxation is 10 seconds.

VIII. RELAXATION OF PREFERENTIAL DISTRIBUTIONS INDUCED BY STRESS

a) GENERAL THEORY

The atoms in a disordered solid solution of cubic structure are distributed isotropically in the absence of an externally applied stress. Thus, in a stress-free solid solution the distribution of solute atoms is isotropic with respect to each individual solute atom, and the nearest solute atom has an equal probability of lying along any of the crystallographically allowed directions. In body-centered cubic (b.c.c.) structures the interstitial positions, namely, the centers of the faces and edges, have tetragonal symmetry, the tetragonal axis pointing along one of the three principal axes. In a stress-free b.c.c. lattice containing interstitial solute atoms, these atoms are distributed isotropically, i.e., the principal axis of the position occupied by a particular solute atom has an equal probability of being parallel to any one of the three principal axes.

The equilibrium distribution of solute atoms is no longer necessarily isotropic in the presence of a stress. Thus in the case of a face-centered cubic (f.c.c.) lattice the axis joining nearest neighbors is along a $<110>$ direction. If the solute atoms are larger than the solvent atoms, two neighboring solute atoms whose axis is along the [110] direction will produce an anisotropic distortion, in the sense that the lattice will be locally extended along the [110] direction more than along the directions normal thereto. A tensile stress along the [110] direction will therefore induce a preferential distribution in which more pairs of solute atoms have their

47. W. S. Gorsky, "Theorie der Ordnungsprozesse und der Diffusion in Mischkristallen von CuAu," *Phys. Zeitschr., Sow.*, VIII (1935), 443.

axes lying along the [110] axis than along any of the other five <110> axes. Similarly, a tensile stress along the [100] axis of a b.c.c. lattice with interstitial solute atoms will induce a preferential equilibrium distribution in which each solute atom has a greater probability of being in an interstitial position whose tetragonal axis is along the [100] axis than along either of the other two <100> axes. The continual striving of the metal to maintain an equilibrium distribution will give rise to anelasticity. In particular, the associated internal friction will be a maximum when the time of relaxation for the establishment of equilibrium is comparable with the period of vibration.

We shall now compute the relaxation strength associated with the establishment of the equilibrium preferential distribution. Toward this end we consider that a tensile stress, σ, is applied along one of the crystallographic axes which may be preferred, e.g., along a <110> axis in a f.c.c. lattice. We shall denote by n the number of elements per unit volume which may acquire a preferential orientation, e.g., number of pairs of solute atoms in the case of substitutional solid solutions, number of solute atoms in the case of interstitial solution in b.c.c. lattices. The quantity n_p will denote the excess number of elements, per unit volume, in the preferred orientation over the number in this orientation when the distribution is random.

The tensile strain, ϵ, and the energy difference, U, of an element in a preferred and a nonpreferred orientation may then be considered as functions of σ and n_p. Thus

$$\epsilon = E_U^{-1}\sigma + \left(\frac{\partial \epsilon}{\partial n_p}\right)_\sigma n_p ,$$

$$U = \left(\frac{\partial U}{\partial \sigma}\right)_{n_p} \sigma + \left(\frac{\partial U}{\partial n_p}\right)_\sigma n_p .$$

Under equilibrium conditions, n_p will be proportional to U; thus

$$n_p = -\frac{\beta n}{kT} U , \tag{189}$$

where β is a numerical coefficient whose precise value will have to be evaluated for each particular case from statistical considerations. The relaxation strength may now be obtained through elimination of n_p and U from the above three equations. We obtain

$$\Delta_E = \frac{-E_U\left(\frac{\partial \epsilon}{\partial n_p}\right)_\sigma \cdot \left(\frac{\partial U}{\partial \sigma}\right)_{n_p}}{\frac{kT}{\beta n} + \left(\frac{\partial U}{\partial n_p}\right)_\sigma} . \tag{190}$$

An estimate may usually be made of the derivative $(\partial\epsilon/\partial n_p)_\sigma$ from experimental data, while no such estimate is available for the derivatives $(\partial U/\partial\sigma)_{n_p}$ and $(\partial U/\partial n_p)_\sigma$. It is therefore desirable to transform these last two derivatives. The first is transformed by forming the perfect differential,

$$d(E-\epsilon\sigma) = -\epsilon d\sigma + U dn_p ,$$

where E is the energy per unit volume. From this equation we obtain

$$\left(\frac{\partial U}{\partial\sigma}\right)_{n_p} = -\left(\frac{\partial\epsilon}{\partial n_p}\right)_\sigma . \tag{191}$$

The second unknown derivative, $(\partial U/\partial n_p)_\sigma$, will be estimated from the following considerations. It is anticipated that both σ and n_p influence U primarily through their effect upon the tensile strain, ϵ. We therefore expect the change in U induced by δn_p to be of the same order of magnitude as the change in U induced by a stress increment $\delta\sigma$, which gives rise to the same strain increment, $\delta\epsilon$, as does δn_p. We therefore set

$$\left(\frac{\partial U}{\partial n_p}\right)_\sigma = \nu\left(\frac{\partial U}{\partial\sigma}\right)_{n_p}\left(\frac{\partial\sigma}{\partial\epsilon}\right)_{n_p}\left(\frac{\partial\epsilon}{\partial n}\right)_\sigma , \tag{192}$$

where ν is a numerical coefficient of the order of magnitude of unity. This numerical coefficient corresponds to the Weiss factor in the theory of the paramagnetism of ferromagnetic materials.

Upon substituting equations (191) and (192) in equation (190), we obtain

$$\Delta_E = \frac{T_0}{T-\nu T_0} , \tag{193}$$

where

$$T_0 = \beta n E_U k^{-1}\left(\frac{\partial\epsilon}{\partial n_p}\right)^2 . \tag{194}$$

This equation for Δ_E is, of course, valid only as long as T is above the critical temperature, T_c, for a self-induced preferential distribution. The relation of T_c to ν and T_0 must be obtained for each particular case through statistical mechanics.

b) EXAMPLE OF SUBSTITUTIONAL SOLID SOLUTION

A relaxation has been observed in α-brass by the author,[48] for which he had no interpretation at the time that the measurements were made. The internal friction of an α-brass single crystal vibrating at a frequency of about 600 cycles/sec was found to rise steadily as the temperature was

48. C. Zener, "Internal Friction of an Alpha Brass Crystal," *Trans. A.I.M.E.*, CLII (1943), 122.

increased to 425° C., and then to decrease as the temperature was further increased. The observations are reproduced in Figure 35. Since zinc atoms are larger than copper atoms, it appears very likely that this anomalous behavior of the α-brass crystal was due to the anelasticity associated with

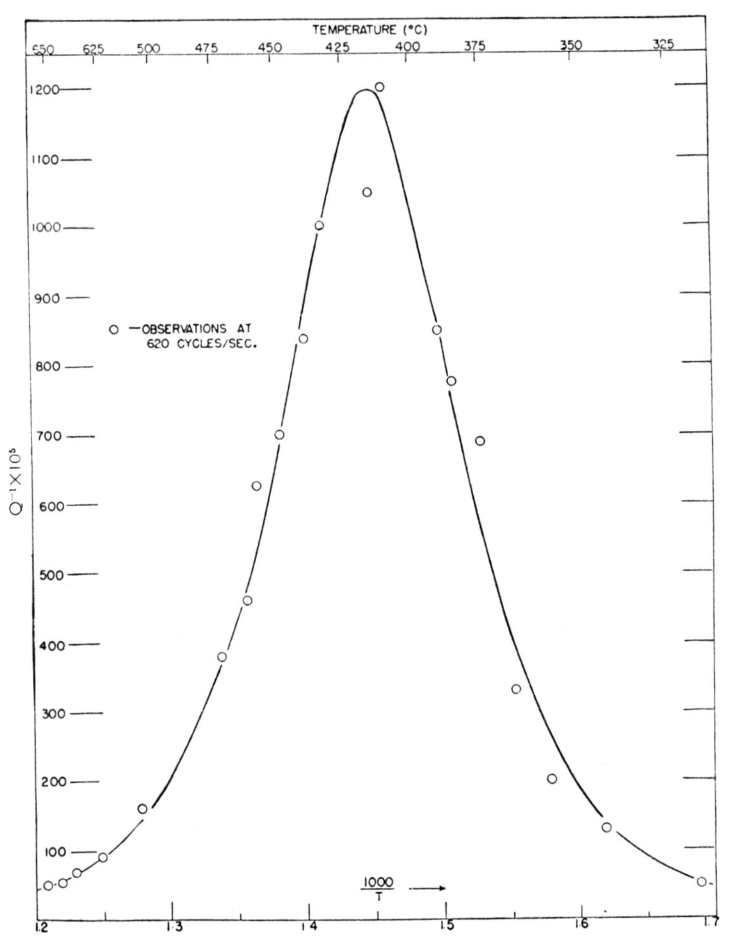

FIG. 35.—Internal friction due to stress induced by preferential orientation of axes joining pairs of solute atoms. (After Zener.)

the relaxation of the preferential distribution induced by the stresses of vibration. The maximum internal friction occurred at that temperature at which the time of relaxation of the preferential distribution was equal to the period of vibration. This viewpoint is further strengthened by the fact that the observations agree very well with a formula for a single time of relaxation with a heat of activation, namely,

$$\tan \delta = \Delta_E \cdot (P + P^{-1})^{-1} , \tag{195}$$

where

$$P = \frac{\tau}{\tau_0} e^{-H/RT} \tag{196}$$

and τ is the period of vibration. The full curve in Figure 35 is the plot of equation (195). The estimated heat of activation H, 33,000 cal/mole, is consistent with the heat of activation for the diffusion of zinc in copper observed by Rhines and Mehl.[49] This heat of activation they found to decrease from 40,000 to 31,000 cal/mole as the zinc concentration increased from 10 to greater than 16 per cent.

In a later paper[50] the author has shown that the above interpretation is consistent with the observed relaxation strength of the α-brass crystal, namely, 0.025 at 400° C. Since the full relaxation curve was observed at only one frequency, it is not possible from these experiments to reach any conclusion regarding the value of the critical temperature, T_c.

It is expected that the same type of anomalous anelastic behavior found for α-brass crystals will be found in all solid-solution crystals in which the solute and solvent atoms have different diameters, as is usually the case. Particularly striking crystallographic effects will be found in body-centered crystals. Here the axes joining closest neighbors will be along the $<111>$ directions. The angles which each of these directions makes with a $<100>$ axis are identical. Hence a tensile stress applied along a $<100>$ direction will not induce a preferential orientation of pairs of solute atoms and hence will not be associated with the above-described type of anelasticity.

c) EXAMPLE OF INTERSTITIAL SOLUTION

The first recorded anomaly in the internal friction of iron as a function of temperature was given by Cantone[51] in 1895. During the ensuing half-century this anomaly was reported by many investigators, in apparent ignorance of the previous work. In Figure 36 is reproduced the observation of Woodruff[52] (1903) upon the influence of temperature on the time of decay of the vibrations in steel tuning forks. Since the time of decay is inversely proportional to the internal friction, the time of decay has a minimum where the internal friction has a maximum. Woodruff's observations were the first to demonstrate that the temperature at which the internal friction is a maximum depends upon the frequency, being higher,

49. F. N. Rhines and R. H. Mehl, "Rates of Solution in the Alpha Solid Solutions of Copper," *Trans. A.I.M.E.*, CXXVIII (1938), 185.

50. C. Zener, "Stress Induced Preferential Orientation of Pairs of Solute Atoms in Metallic Solid Solution," *Phys. Rev.*, LXXI (1947), 34.

51. M. Cantone, *Nuovo cimento*, I (1895), 165, 205; IV (1896), 270, 354.

52. E. Woodruff, "A Study of the Effects of Temperature upon a Tuning Fork," *Phys. Rev.*, XVI (1903), 321.

the higher the frequency. A very extensive study was made by Robin[53] in 1910–11, typical results being reproduced in Figure 37. Robin was the first to demonstrate that the anomalous internal friction was greater in low- than in high-carbon steel and that the anomaly disappears upon plastic deformation, reappearing again after annealing. Several later authors have reported less detailed observations upon the anomaly.[54]

Mild steel also shows an anomalous magnetic aftereffect which cannot be understood from magnetic considerations alone, as was first demon-

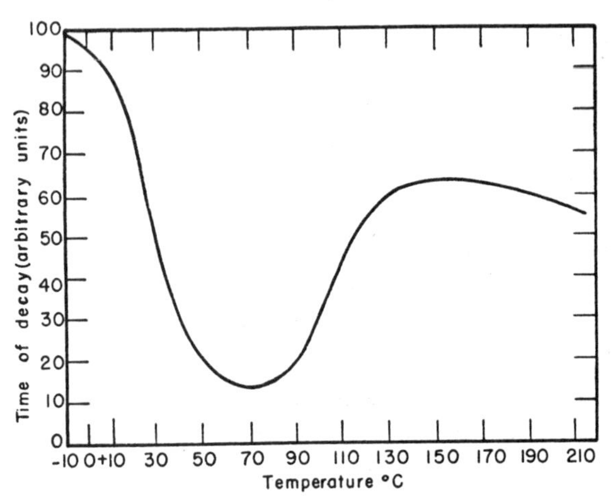

Fig. 36.—Internal friction of steel tuning forks. (After Woodruff.) f = 128 cycles/sec

strated by Ewing (1885).[55] In careful observations upon both the magnetic and the mechanical aftereffect in carbonyl iron, Richter,[56] demonstrated the very close relation between the two effects. An example of this relation is illustrated in Figure 38. Richter demonstrated that the magnetic and mechanical aftereffects have the same heat of activation, namely, 19,600 cal/mole, and that the magnitude of both aftereffects has a temperature coefficient of -0.003 near room temperature, a temperature coefficient

53. F. Robin, "Report on the Wear of Steels and on Their Resistance to Crushing," *Iron and Steel Inst., Carnegie Schol. Mem.*, II (1910), 259; "Phénomène de l'extinction de son dans le fer," *Compt. rend., Acad. Sci.*, CL (1910), 780; and "The Variations in the Acoustical Properties of Steel with Changes in Temperature," *Iron and Steel Inst. Carnegie Schol. Mem.*, III (1911), 125.

54. K. Jokibe and S. Sakui, *Phil. Mag.*, XLII (1921), 397; A. Esau and E. Voigt, *Zeitschr. f. tech. Phys.*, XI (1930), 78.

55. J. Ewing, *Phil. Trans. Roy. Soc., London,* CLXXVI (1885), 554.

56. G. Richter, "Über die mechanische und magnetische Nachwirkung des Carbonyleisens," *Ann. d. Phys.*, XXXII (1938), 683; and chapter on "Magnetische und mechanische Nachwirkung," in R. Becker (ed.), *Probleme der technischen Magnetisierungskurve* (Berlin: Springer, 1938; New York: Edwards Bros., 1944).

which is consistent with a variation as $1/T$. Of great significance was Richter's observation that a strong magnetic field suppressed the magnetic aftereffects but had no influence upon the mechanical aftereffects. It was therefore evident that their common origin was of a mechanical rather than of a magnetic nature. This conclusion invalidated Snoek's original hypothesis[57] that the elastic aftereffect had its origin in magnetostriction and in the concentration of solute atoms along the boundaries of the magnetic domains. The essential role that interstitial solute atoms play in the aftereffects was found by Snoek[58] when he removed all traces of carbon and of nitrogen, thereby removing both the magnetic and the elastic after-

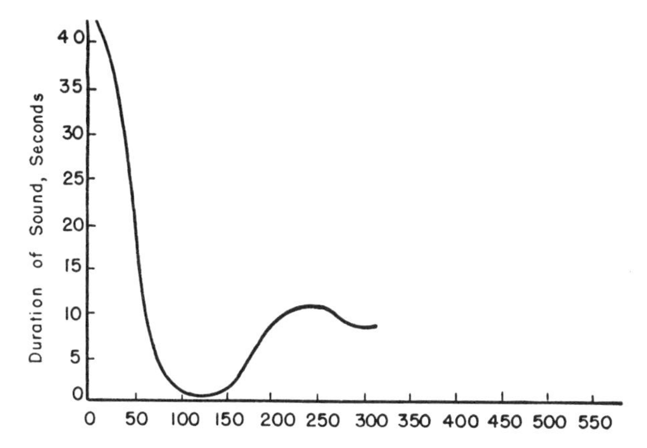

FIG. 37.—Duration of sound in annealed carbon steels. (After Robin)

effects. When less than 0.01 per cent of either carbon or nitrogen was replaced, the aftereffects returned. Two years later he[59] developed the theory of the preferential distribution of interstitial carbon and nitrogen atoms.

In his theory Snoek assumes that the interstitial positions occupied by iron and nitrogen atoms are at the centers of the cube faces and at the centers of the cube edges, which are crystallographically equivalent. This assumption as to the location of the dissolved carbon and nitrogen atoms is consistent with X-ray evidence,[60] but is inconsistent with the viewpoint that interstitial solute atoms must be located at the largest interstitial

57. J. Snoek, "Time Effects in Magnetisation," *Physica*, V (1938), 663.

58. J. Snoek, "Mechanical After-effect and Chemical Constitution," *Physica*, VI (1939), 591.

59. J. Snoek, *Physica*, VIII (1941), 711; "Tetragonal Martensite and Elastic After-effect in Iron," *ibid.*, IX (1942), 862; and *Chemisch Weekblad*, XXXIX (1942), 454.

60. N. J. Petch, *Jour. Iron and Steel Inst.*, No. 1, 1943, p. 221; H. Lipson and Audrey M. Parker, *Jour. Iron and Steel Inst.*, No. 1, 1944, p. 123.

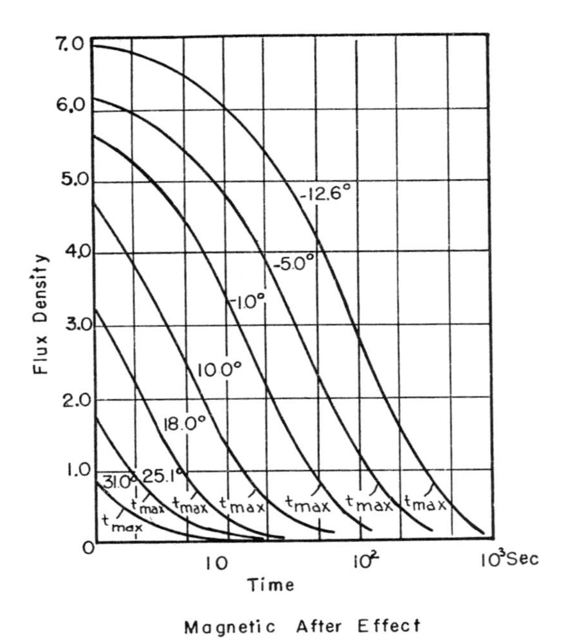

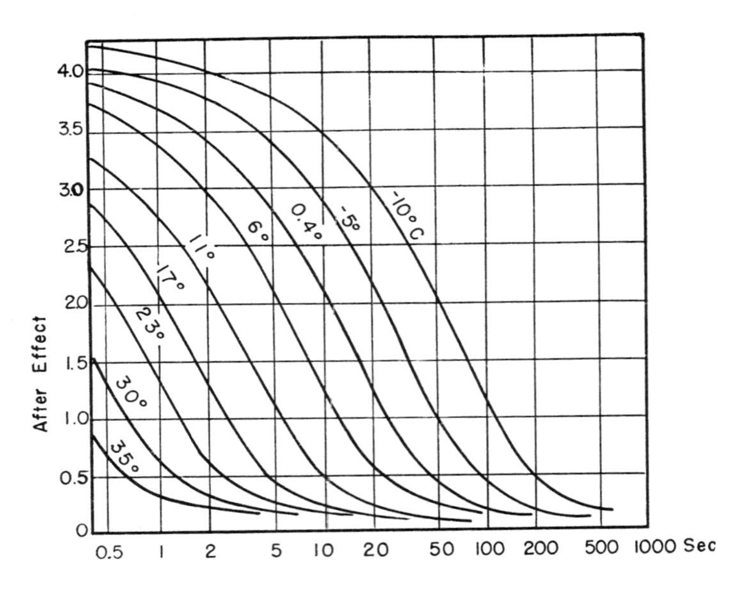

FIG. 38.—Comparison of magnetic and mechanical aftereffect in carbonyl iron. (After Richter)

holes. Thus the interstitial position at $(\frac{1}{2}, \frac{1}{4}, 0)$ corresponds to a considerably larger hole than the position $(\frac{1}{2}, 0, 0)$ assumed by Snoek. When the criterion of minimum strain energy is employed to decide the location of the interstitial solute atoms, it is found that the $(\frac{1}{2}, 0, 0)$ type positions are indeed correct. Although the iron lattice is cubic, the local symmetry of these interstitial positions is tetragonal. Thus the interstitial position $(\frac{1}{2}, 0, 0)$ is closer to the two neighboring carbon atoms along the [100] axis than to the four other neighboring carbon atoms. The [100] direction may therefore be regarded as the tetragonal axis of the $(\frac{1}{2}, 0, 0)$ interstitial position. Similarly, every interstitial position has associated with it a tetragonal axis which is parallel to one of the three principal $<100>$ axes. When an iron crystal is under no stress, all interstitial positions are equivalent, and the interstitial solute atoms are therefore randomly distributed. When, however, a tensile stress is applied along a principal axis, the equilibrium distribution will be one in which those interstitial positions are preferred whose tetragonal axes are parallel to the tensile axis, since an interstitial solute atom distorts the surrounding lattice in an asymmetrical manner, causing a greater dilation along its tetragonal axis than normal to this axis. Conversely, a tensile stress along the tetragonal axis will tend to induce a preferential distribution of interstitial solute atoms in those interstitial positions with tetragonal axes parallel to the tensile axis.

The Snoek theory of the anomalous behavior of mild steel is thus a special case of the general theory developed in section *a* (pp. 111–13). It is therefore anticipated that the relaxation strength will be given by equations (193) and (194). In applying these equations, we shall initially assume that the tensile stress is along the [100] axis. When we take cognizance of the fact that there are twice as many nonpreferred as preferred positions, we find the numerical coefficient β, defined in equation (189), to be (2/9). It remains to evaluate T_0. This quantity may be obtained from the data relating changes in the lattice parameters of martensite with carbon composition.[61] From these data one deduces that

$$\left(\frac{\partial e}{\partial C_p}\right)_\sigma = 1.0\,, \tag{197}$$

where δC_p is the change in concentration, referred to total number of iron atoms, of the carbon atoms in the preferred interstitial positions, $-\delta C_p$ the change in the nonpreferred positions. Using these data, one obtains

$$T_0 = 1190 X_c\,,$$

61. S. Epstein, *The Alloys of Iron and Carbon*, Vol. I: *Constitution* (New York: McGraw-Hill Book Co., 1936), p. 212.

where X_c is the concentration of carbon atoms, in weight per cent. For concentrations of carbon obtainable in solid solution in mild steel, T_0 is very small, less than 50° K. Equation (193) thus reduces to

$$\Delta_E = \frac{1190 X_c}{T}. \tag{198}$$

As an example, for iron with 0.01 weight per cent of carbon, the above formula gives the value of Δ_E at room temperature as 0.040. The experimental value of Δ_E for these conditions has been found by Dijkstra and reported by Polder[62] as 0.043. The verification of the temperature dependence of Δ_E as given by equation (198) is contained in the work of Richter.[63] The most convincing verification of the Snoek theory has been given by Dijkstra[64] through his observations upon the crystallographic dependence of the relaxation strength in single crystals. If a tensile stress is applied along a <111> axis, no interstitial positions become preferred, since all the tetragonal axes are equally inclined toward the tensile axis. One should therefore anticipate no relaxation due to carbon or nitrogen atoms. This has been shown by Dijkstra to be indeed the case.

It would be anticipated that the anelastic effects associated with carbon and nitrogen in iron would be associated with all interstitial solutions in b.c.c. metals, provided that the solute atoms are sufficiently large to prefer the $(\frac{1}{2}, 0, 0)$ type positions over the $(\frac{1}{2}, \frac{1}{4}, 0)$ type positions. Such effects have been described in detail by Kê[64a] for the case of carbon, nitrogen, and oxygen dissolved in tantalum, and have likewise been observed in columbium[64b] and in tungsten.[64c]

Recent observations by Dijkstra[65] indicate the potential usefulness of anelastic measurements in studying the metallurgy of interstitial solutions in b.c.c. metals. Since the relaxation strength, or any quantity related thereto, such as the internal friction at the optimum temperature, is a measure of the concentration of solute atoms in solid solution, its measurement may be used to follow the rate at which the solute atoms come out of solution and, finally, even to measure the solution pressure in equilib-

62. D. Polder, "Theory of the Elastic After-effect and the Diffusion of Carbon in Alpha-Iron," *Philips Research Repts.*, I (1945), 1.

63. *Loc. cit.*

64. L. J. Dijkstra, "Elastic Relaxation and Some Other Properties of the Solution of Carbon and Nitrogen in Iron," *Philips Research Repts.*, II (1947), 357.

64a. T. S. Kê, "Internal Friction in the Interstitial Solid Solutions of C and O in Tantalum," *Phys. Rev.*, 1948.

64b. L. J. Dijkstra, personal communication.

64c. J. Snoek, personal communication.

65. L. J. Dijkstra, "Precipitation Phenomena in the Solid Solutions of Nitrogen and Carbon in Alpha Iron below the Entectoid Temperature," *Trans. A.I.M.E.*, CLXXXV (1949), 252.

rium with the precipitate. In such a study of nitrogen in iron Dijkstra has not only measured with a precision heretofore unattainable the solution pressure of iron nitride from the eutectoid temperature down to 200° C. but has also discovered a new phase and has measured its solution pressure, a phase which is metastable with respect to iron nitride. The structure of this metastable phase has later been identified by Jack.[65a]

In our prior analysis it has been assumed that the applied stresses are so low that the degree of preferential distribution is a linear function of stress. This restriction will now be removed. If n is the total number of interstitial atoms per unit volume and n_p is the number in the interstitial positions with tetragonal axes parallel to the local tensile axes, we may define the order parameter, P, by

$$P = \tfrac{1}{2}\left(3\frac{n_p}{n} - 1\right). \tag{199}$$

When defined in this manner, P varies from 0 to 1 as the distribution changes from complete randomness to preferred positions for all the interstitial atoms. If U denotes the decrease in potential energy which an interstitial atom suffers in changing from a nonpreferred to a preferred interstitial position, then it may readily be deduced that

$$P = \frac{e^x - 1}{e^x + 2}, \tag{200}$$

where

$$x = \frac{U}{kT}. \tag{201}$$

It now remains to deduce how the energy change, U, depends on the stress. We assume that U is independent of temperature. Then, at the absolute zero temperature,

$$d(U - e\sigma) = -\, e d\sigma - U dn_p\,, \tag{202}$$

where U is the energy per unit volume. Since the left-hand side is a perfect differential,

$$\left(\frac{\partial U}{\partial \sigma}\right)_{n_p} = \left(\frac{\partial e}{\partial n_p}\right)_{\sigma}. \tag{203}$$

If we now assume that U is a linear function of stress, we obtain

$$U = \frac{V}{N}\frac{\partial e}{\partial C_p}\sigma\,. \tag{204}$$

65a. K. H. Jack, *Acta Cryst.*, III (1950), 392; *Proc. Roy. Soc. London*, CCVIII (1951), 216.

In the case of carbon, we obtain, in view of equation (197),

$$x = \frac{VE}{RT}\, e\,,$$

where e is the elastic strain. At room temperature this equation becomes

$$x = 386\, e\,.$$

In Figure 39 the degree of order, P, is plotted as a function of x and of e. It is seen that a large reduction of the anelastic effects by residual stresses

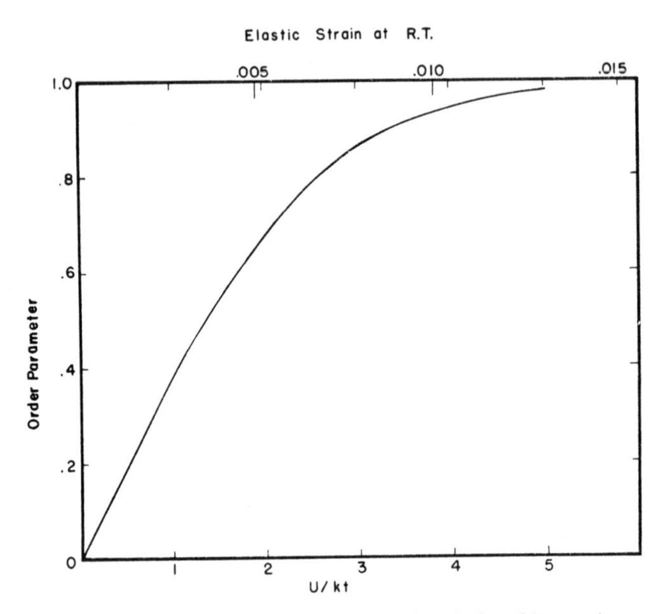

FIG. 39.—Degree of order of carbon in iron induced by strain

would require these stresses to give an elastic strain of about 0.01, that is, stresses of the magnitude of 300,000 psi.

It has previously been mentioned that in mild steel the concentration of carbon in solid solution is always so small that T_0 and hence also probably T_c are less than 50° K. In mild steel we may therefore neglect the interaction of the carbon atoms upon one another. The author[66] has previously suggested that it is just this interaction which gives rise to the tetragonality of quenched steel, a body-centered structure in which all the carbon has been retained in solution, a structure known as "martensite." It looks as though in martensite the critical temperature for self-induced preferential distribution, T_c, is above room temperature. It is therefore of impor-

66. C. Zener, "Kinetics of Decomposition of Austenite," *Trans. A.I.M.E.*, CLXVII (1946), 550.

tance to obtain the relation between T_c and ν and T_0. Toward this end we write the expression for the free energy, F, of a unit volume of unstressed iron. Upon letting M denote the total number of interstitial positions, n the number of carbon atoms, n_p the number in the preferred type in positions, and, finally, Δn_p the excess of n_p over $n/3$, we obtain

$$F = \frac{1}{2} \frac{\partial U}{\partial n_p} (\Delta n_p)^2 - kT \ln \left\{ \frac{M!}{n_p! (M - n_p)!} \cdot \frac{(2M)!}{(n - n_p)! (2M - n + n_p)} \right\}.$$

The equilibrium value of Δn_p is determined by the condition

$$\frac{\partial F}{\partial \Delta n_p} = 0 .$$

When, as is always the case in actual specimens, the total number of carbon atoms is very small compared with the number of iron atoms, i.e., $n/M \ll 1$, this equilibrium condition reduces to

$$-n \left(\frac{\partial U}{\partial n_p} \right)_\sigma x = kT \ln \left(\frac{1 + 3x}{1 - \frac{3x}{2}} \right), \tag{205}$$

where

$$x = \frac{\Delta n_p}{n} . \tag{206}$$

When T is sufficiently large, equation (205) has no solution other than $x = 0$. As T is decreased, this equation attains two additional roots. The critical temperature, T_c, is determined by the condition that at T_c the transition from the random distribution to the preferred distribution is attended by no change in free energy. This equilibrium condition is satisfied when the integral of the left member of equation (205) with respect to x from $x = 0$ to the third root is equal to the corresponding integral of the right member. This root cannot be found graphically merely by plotting the right member of equation (205). As shown by Figure 40, such a plot does not have sufficient dispersion in the region of interest. Upon taking cognizance of the fact that areas remain unaltered by simple shears, we see that the requisite dispersion can be obtained by a suitable simple shear parallel to the vertical axis. Such a sheared plot is shown in Figure 41, from which plot a graphical solution for x_c gives

$$x_c = 0.34 ,$$

$$-n \frac{\left(\frac{\partial U}{\partial n_p} \right)_\sigma}{kT_c} = 4.02 . \tag{207}$$

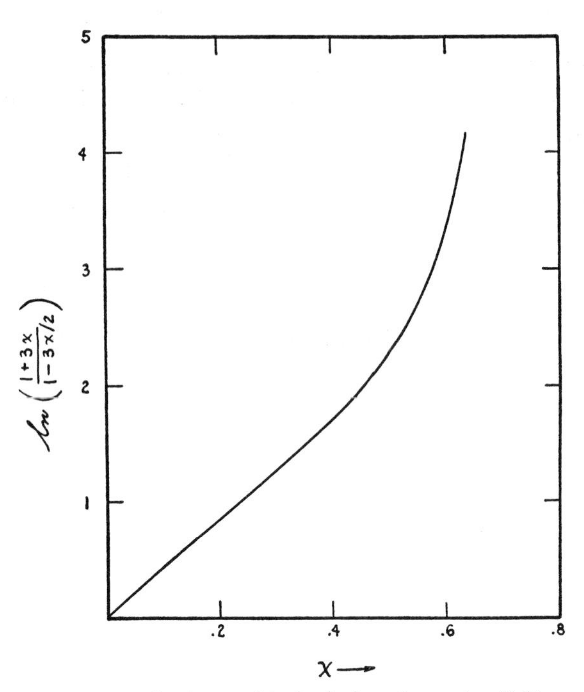

FIG. 40.—Plot for graphical solution of equation (167)

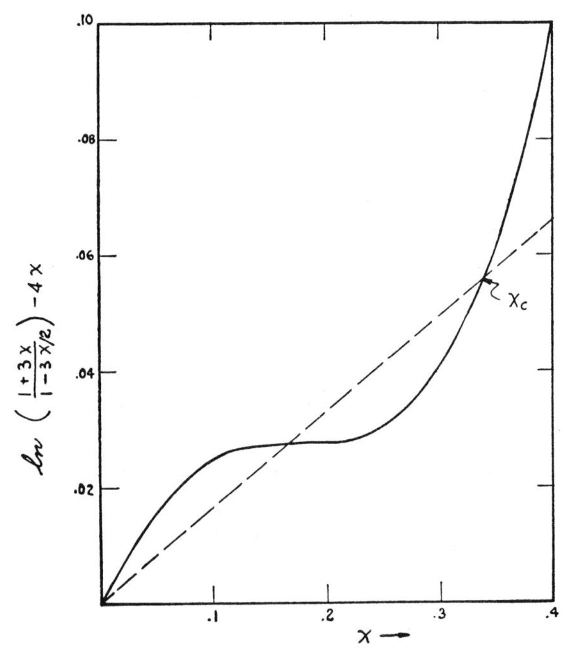

FIG. 41.—Figure 40 after being subjected to a suitable simple shear parallel to vertical axis.

Upon combining equations (192), (194), and (207) and upon using our previously found value of $\frac{2}{9}$ for β, we obtain

$$T_c = 1.05\nu T_0 . \tag{208}$$

Upon returning to equation (193), we see that the maximum theoretical value of the relaxation strength, namely, the value at T_c, is given by

$$(\Delta_E)_{\max} = \frac{20}{\nu} .$$

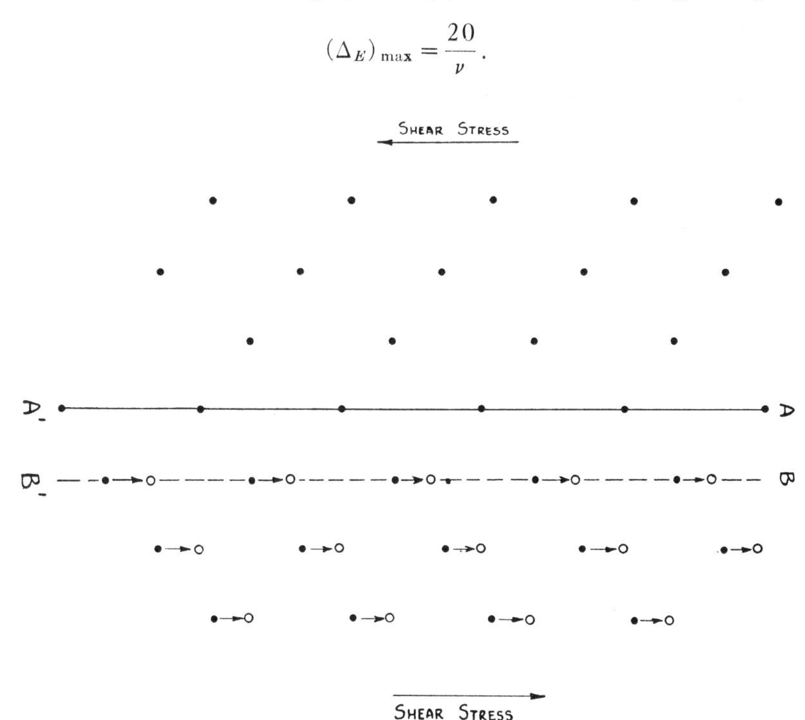

FIG. 42.—Illustration of movement of twin interface. Twin interface is a (011) plane of tetragonal lattice.

Below the critical temperature, relaxation will come primarily through the movement of interfaces between regions in which the preferred directions for the tetragonal interstitial positions are along different principal axes. A possible example of such an interface is illustrated in Figure 42, where the interface is in a (011) plane. Residual stresses will impede the movement of these interfaces in a manner analogous to that in which they impede the movement of the boundaries of magnetic domains.

The above interpretation of martensite as an ordered structure is based upon the assumption of an interaction between the solute carbon atoms of sufficient magnitude to induce an ordered distribution. A detailed study by Kê[66a] has shown that such interaction is extremely small in the case of oxygen dissolved interstitially in tantalum.

B. INHOMOGENEOUS RELAXATION

I. CONCEPT OF TWO-COMPONENT SYSTEMS

In the previous section we considered that in every element of the specimen stress can relax through the relaxation of the fluctuations of some thermodynamic potential, such as temperature. It has long been recognized[67] and has been repeatedly discussed[68] that anelastic phenomena may also arise through a mixture of two phases, one of a truly viscous nature, in the sense that therein

$$\frac{d}{dt}(\text{shear strain}) = G^{-1}\left\{\frac{d}{dt}(\text{shear stress}) + \tau^{-1}\,\text{shear stress}\right\}, \qquad (209)$$

and one of a truly elastic nature, in the sense that

$$\text{Strain} \sim \text{Stress}. \qquad (210)$$

The coexistence of both types of regions is necessary in order to interpret the observed mechanical behavior of metals. If only the first type of media were present, a specimen would behave like a supercooled fluid, creeping but manifesting no creep recovery.

The qualitative behavior of a two-component system may readily be understood by reference to Figure 43. The specimen with no load and in a completely relaxed state is illustrated in *a*. The region within the rectangle is considered as viscous and therefore obeys equation (209). The region outside the rectangle is perfectly elastic and therefore obeys equation (210). Let us now consider that an external load is suddenly applied. With no essential loss of generality, we may regard the instantaneous (i.e., the unrelaxed) modulus of the viscous region as identical with that of the surrounding elastic medium. Just after the load is applied, the deformation is then everywhere uniform, as illustrated in *b*. The shear stress within the viscous region at once starts to relax. If the elastic matrix remained rigid during this relaxation, the shear stress in the viscous region would relax with the time of relaxation, τ, in equation (209). The surrounding elastic matrix will not, however, remain rigid during this relaxation. The equa-

66a. T. S. Kê, "Stress Relaxation by Interstitial Atomic Diffusion in Tantalum," *Phys. Rev.* (1948).

67. C. Maxwell, *Scientific Papers* (Cambridge, 1890), II, 623; J. H. Poynting and J. J. Thomson, *A Text Book of Physics—Properties of Matter* (London: Griffin, 1907), p. 57.

68. For review, see chap. vi of Fromm's article on "Nachwirkung und Hysteresis" in Auerbach-Hort's *Handbuch der physikalischen und technischen Physik* (Leipzig: Johann Barth, 1931), IV, 1; 521–29.

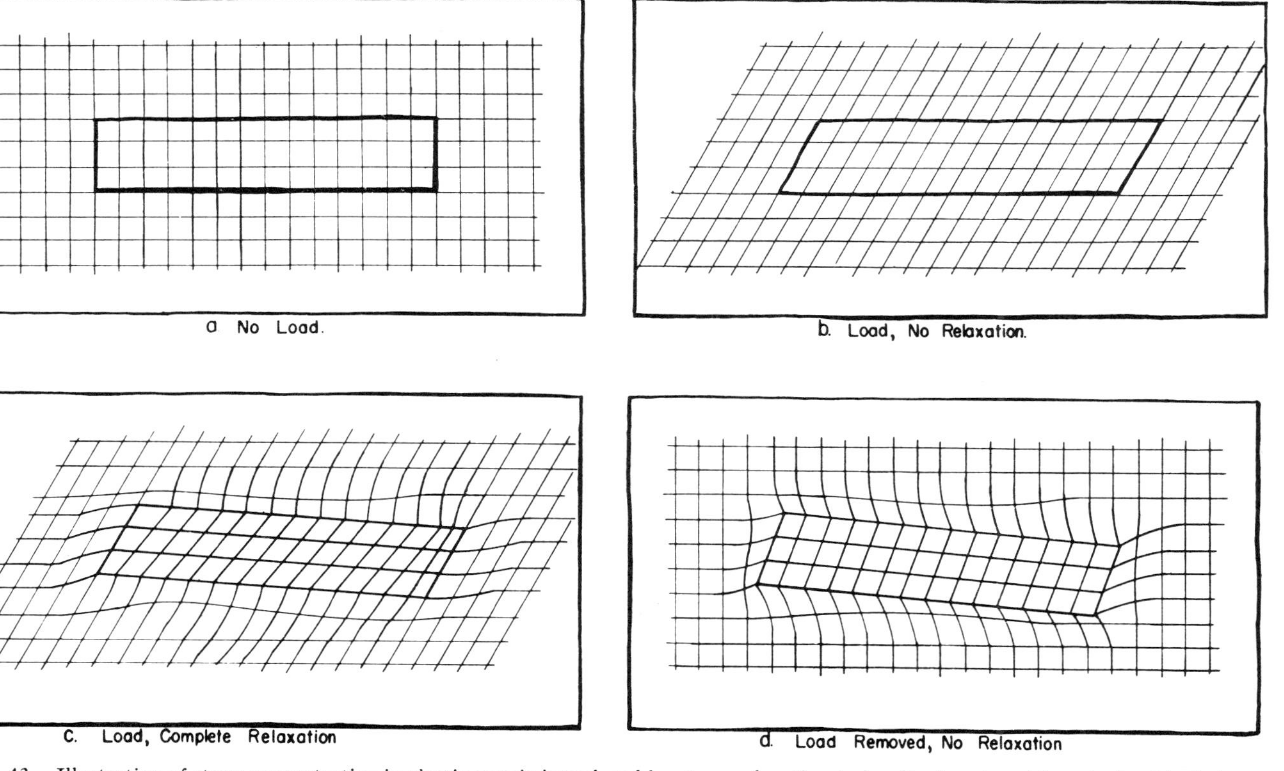

FIG. 43.—Illustration of stress concentration in elastic matrix introduced by stress relaxation in localized regions (*c*), and of residual stresses which remain after external load is removed (*d*).

tions of equilibrium require that the shear stress in the surrounding elastic matrix suffer at least a partial relaxation, accompanied by shear strains, as illustrated in *c*. This relaxation in the elastic matrix results in an over-all slow deformation, or creep. Let us now consider that the load is removed. The instantaneous strain which accompanies this removal is uniform throughout. Far away from the viscous region the elastic matrix will be under essentially zero stress. In the immediate vicinity of the viscous region the elastic matrix will have, however, residual stresses due to the strain which has remained in the viscous region, as illustrated in *d*. These residual stresses, in turn, cause shear stresses in the viscous region. The gradual relaxation of these latter shear stresses ultimately brings about complete relaxation of all stresses, and the specimen finds itself in its initial state as illustrated by *a*. It is this relaxation of all residual stresses which gives rise to the elastic aftereffect.

One remarkable feature of the two-component system is the large magnitude of the anelastic effects which may be produced by an almost negligible amount of viscous material. Thus suppose that the viscous phase is distributed in the form of a thin disk. As already explained, shear stress is partially relaxed in the elastic matrix for some distance on either side of the disk. The distance over which this relaxation extends is essentially independent of the thickness of the viscous disk and will be comparable with the radius of the disk. If the disk is several atomic diameters wide and has a radius of several hundred atomic diameters, shear stress will be relieved in a volume of the order of magnitude of a hundred times as large as the viscous region itself. The concept of a two-component system therefore requires only a negligible amount of the viscous phase.

A second remarkable feature of the two-component system is the building-up of high stress concentrations within the elastic matrix as the stress relaxes within the viscous region. The formation of such stress concentrations may readily be understood by reference to Figure 43. While the shear stress relaxes in the elastic matrix on either side of the viscous region, high stress concentrations are built up near the ends of the viscous region. The necessity for such stress concentrations may be seen also from general energy considerations. If L is the applied load, the total strain energy stored in the specimen before and after stress relaxation is proportional to L^2/M_U and L^2/M_R, respectively, where M_U and M_R are the unrelaxed and relaxed moduli. Since M_R is less than M_U, the strain energy stored in the specimen is greater after relaxation than before. Since, however, some regions have suffered a relaxation of stress, other regions must be under a correspondingly greater stress. An estimate of the order of magnitude of the stress concentration may be obtained from the computations of Inglis[69] upon the shear stress computations due to cracks. Thus, suppose that the viscous region is in the shape of a thin disk and that the applied forces result in shear stresses in the plane of the disk, with no normal tensile stresses. The strains in the surrounding elastic region in the completely relaxed condition will then be precisely the same as if the viscous region had been a cracklike cavity, for which Inglis made his computations. His results may be expressed most conveniently in terms of the radius of the disk, a, and in terms of the radius of curvature, r, of the edge of the crack. His results are

$$\text{Maximum tensile stress at edge of crack} = \left(\sqrt{\frac{r}{a}} + \sqrt{\frac{a}{r}}\right) \times \text{(over-all shear stress)}\,.$$

Thus, if the radius of curvature at the edge of the disk is only one-hundredth the radius of the disk, stress relaxation within the viscous region develops a tensile stress within the elastic matrix ten times as large as the over-all shear stress.

The stress concentrations in the elastic matrix developed by stress relaxation in the viscous regions may play an important role in the fracture of metals. It has long been recognized that the observed over-all fracture strength of metals is at least an order of magnitude less than the value computed for a perfect lattice. The theoretical value of the tensile stress at fracture, T, may be expressed as

$$T = e_0 E\,,$$

where e_0 is the elastic strain at fracture and E is Young's modulus. The theoretical value of e_0 for crystals is larger than 0.1; thus for NaCl[70] it is 0.15. On the other hand, the elastic strain at fracture is never so large as 0.01. One is tempted to attribute the discrepancy between the observed and the theoretical fracture stress to the presence of small microcracks, which, through their stress concentrations, locally raise the tensile stress to the theoretical value for fracture. In fact, many fracture phenomena may be understood in terms of the presence of such microcracks.[71] As

69. C. E. Inglis, "Stresses in a Plate Due to the Presence of Cracks and Sharp Corners," *Trans. Inst. Naval Architects*, LV, Part I (1913), 219.

70. F. Zwicky, *Phys. Zeitschr.*, XXIV (1923), 131.

71. C. Zener and J. H. Hollomon, "Plastic Flow and Rupture of Metals," *Trans. A.S.M.*, XXXIII (1944), 163; and J. H. Hollomon, *The Problem of Fracture* (American Welding Soc., 1946).

first pointed out by Griffith,[72] a microcrack requires a critical over-all tensile stress before it will spread, the magnitude of this critical stress increasing as the radius of the crack decreases. This relation may be understood from two viewpoints, from surface energy considerations and from stress concentration considerations. According to the first viewpoint, we inquire as to whether an increase in the radius of the crack will result in a lowering or a raising of the energy of the system. Two types of energy must be considered, elastic strain energy and surface energy. If the tensile stress is normal to the plane of the crack and the radius of the crack is a, stress will be relieved in a volume of the order of magnitude of $(4\pi/3)a^3$. If the over-all strain is maintained constant, the strain energy associated with the crack is therefore

$$\text{Strain energy} = -C\,\frac{4\pi a^3}{3}\cdot\frac{\frac{1}{2}T^2}{E},$$

where C is a numerical constant of the order of magnitude of unity and T is the over-all tensile stress. The surface energy associated with the crack is

$$\text{Surface energy} = 2\pi a^2 S\ ,$$

where S is the energy per unit surface, i.e., the surface tension. The crack will spread under the influence of the tensile stress only if the decrease of strain energy resulting therefrom is greater than the increase in surface energy. The critical tensile stress is obtained by equating to zero the derivative of the total energy with respect to the radius a. One thereby obtains

$$T = \sqrt{\frac{2SE}{Ca}}\,. \tag{211}$$

According to the second viewpoint, a crack will propagate when its stress concentration locally raises the over-all tensile stress to the theoretical fracture stress. Since the stress concentration factor[73] is $2\sqrt{a/r}$, where r is the radius of curvature of the edge of the crack, this second viewpoint gives

$$T = \tfrac{1}{2}\sqrt{\frac{r}{a}}\;e_0 E\,. \tag{212}$$

According to both viewpoints, T varies inversely as the square root of a. In order that equations (211) and (212) may be compatible, it is necessary that

$$2S = \tfrac{1}{2}C\cdot\tfrac{1}{2}E\,e_0^2\cdot r\ .$$

72. A. A. Griffith, "The Phenomena of Rupture and Flow in Solids," *Phil. Trans., Roy. Soc., London*, CCXXI (1920), 163; and "Theory of Rupture," *Proc. First Internat. Cong. Appl. Mech., Delft, 1924.*

73. Inglis, *op. cit.*

The left-hand side is the work done in opening up a crack of unit area. When we take r as one interatomic distance, it is clear that this equation will be satisfied by a numerical constant, C, of the order of magnitude of unity. Griffith's concept of a critical tensile stress for the propagation of a crack poses the question as to the origin of the original microcracks. Their origin very probably lies in the stress concentrations developed by stress relaxation within amorphous regions. The role of stress relaxation in opening up microcracks will be further discussed in subsequent sections, where the different types of viscous regions are treated. The development of microcracks by localized stress relaxation has been considered by Murgatroyd[74] as the origin of stress fatigue in glass, i.e., fracture by a sustained load.

A third remarkable feature of the two-component system is the wide variety of types of relaxation spectra which it may have. If all the localized viscous regions had essentially the same size and shape, the relaxation spectrum could be described essentially by a single time of relaxation, as is frequently the case for the relaxation of thermodynamic potentials. A review of past experimental work upon anelasticity presumably arising from viscous regions is given by Fromm.[75] In no cases did the observations indicate a relaxation spectrum even approaching that for a single relaxation time. In fact, all observations indicate a relaxation spectrum whose intensity continually rises or remains stationary with increasing times of relaxation. No cases have been observed in which the intensity of the relaxation spectrum decreases with increasing times of relaxation. These observations may be interpreted in several ways.

As we have already mentioned, a concentrated relaxation spectrum would be observed if all the localized viscous regions were of the same shape and size. One might interpret the observed anelastic behavior in terms of a distribution of size and shape of the viscous regions. Thus suppose that the regions were disk-shaped and that all had the same thickness but various radii. Then the total shear strain for complete stress relaxation would be larger in the disks of greater radii, and hence the larger viscous regions would have longer relaxation times.

On the other hand, the viscous phase might be regarded as forming a continuous network which incloses the elastic matrix. If this network were in the form of regular polyhedra, then the edges and corners of the polyhedra would block viscous deformation, the viscous deformation be-

74. J. B. Murgatroyd, *Jour. Soc. Glass Technol.*, XXVIII (1944), 406; *Nature*, CLIV (1944), 51; J. B. Murgatroyd and R. F. Sykes, *Nature*, CLVI (1945), 717.

75. *Op. cit.*

ing thereby confined to the faces of the polyhedra. If the polyhedra were all of the same size, their relaxation times would all be equal, and the relaxation spectrum would be concentrated. On the other hand, a wide distribution in sizes of the polyhedra would give rise to a wide relaxation spectrum. Such a viscous network could give rise to a wide relaxation spectrum in still another manner. If the network could be represented by regular polyhedra, viscous flow might extend over wide areas unimpeded by edges, while viscous flow in other areas might be limited by closely spaced edges.

The form of the distribution function, $\Delta(\tau)$, and its methods of determination have recently become a subject of intense concern among those interested in the mechanical properties of high polymers. The activity[76] in this field during 1944–46 is in marked contrast to the almost complete lack of work in the corresponding field of metals.

II. STRESS RELAXATION ALONG PREVIOUSLY FORMED SLIP BANDS

a) ROLE OF SLIP BANDS AND DISLOCATIONS IN PLASTIC DEFORMATION

That plastic deformation in metals occurs within isolated regions, *slip bands*, rather than homogeneously throughout the material, was first clearly recognized by Ewing and Rosenhain.[77] Very careful examination

76. T. Alfrey, "A Molecular Theory of the Viscoelastic Behavior of an Amorphous Linear Polymer," *Jour. Chem. Phys.*, XII (1944), 374; "Non-homogeneous Stresses in Viscoelastic Media," *Quart. Jour. Appl. Math.*, II (1944), 113; and *The Mechanical Behavior of High Polymers* (Inter-Science Co., 1944); H. A. Braendle and W. B. Wiegand, "GR-S, an Elastically Inverted Polymer," *Jour. Appl. Phys.*, XV (1944), 304; F. S. Conant and J. W. Liska, "Some Low Temperature Properties of Elastomers," *Jour. Appl. Phys.*, XV (1944), 767; J. H. Dillon, I. B. Prettyman, and C. I. Hall, "Hysteretic and Elastic Properties of Rubberlike Materials under Dynamic Shear Stresses," *Jour. Appl. Phys.*, XV (1944), 309; P. Flory, "Network Structure and the Elastic Properties of Vulcanised Rubber," *Chem. Rev.*, XXXV (1944), 51; H. M. James and E. Guth, "Theory of the Elasticity of Rubber," *Jour. Appl. Phys.*, XV (1944), 294; M. Mooney, W. E. Wolstenholm, and D. S. Villars, "Drift and Relaxation of Rubber," *Jour. Appl. Phys.*, XV (1944), 324; A. V. Tobolsky, I. B. Prettyman, and J. H. Dillon, "Stress Relaxation of Natural and Synthetic Rubber Stock," *Jour. Appl. Phys.*, XV (1944), 380; T. Alfrey and P. Doty, "Methods of Specifying the Properties of Viscoelastic Materials," *Jour. Appl. Phys.*, XVI (1945), 700; R. D. Andrews, R. B. Mesrobian, and A. V. Tobolsky, "Creep and Relaxation in Rubber Products at Elevated Temperatures," *Trans. A.S.M.E.*, LXVII (1945), 569; H. E. Greene and D. L. Loughborough, "Some Physical Properties of Elastomers at Low Temperature," *Jour. Appl. Phys.*, XVI (1945), 3; A. V. Tobolsky and R. D. Andrews, "Systems Manifesting Superposed Elastic and Viscous Behavior," *Jour. Chem. Phys.*, XIII (1945), 3; T. Alfrey, "Methods of Representing the Properties of Viscoelastic Materials," *Quart. Jour. Appl. Math.*, III (1945–46), 143; J. Blatz and A. V. Tobolsky, "Creep under Constant Load in Tension," *Jour. Chem. Phys.*, XIV (1946), 113; S. L. Dart and E. Guth, "Elastic Properties of Cork. I. Stress Relaxation of Compressed Cork," *Jour. Appl. Phys.*, XVII (1946), 314; M. S. Green and A. V. Tobolsky, "A New Approach to the Theory of Relaxing Polymeric Media," *Jour. Chem. Phys.*, XIV (1946), 80; W. J. Lyons, "The General Relations for Flow in Solids and Their Applications to the Plastic Behavior of Tire Cords," *Jour. Appl. Phys.*, XVII (1946), 472; G. Kirkwood, "Elastic Loss and Relaxation Times in Cross-linked Polymers," *Jour. Chem. Phys.*, XIV (1946), 51; M. D. Stern and A. V. Tobolsky, "Stress-Time-Temperature Relation in Polysulfide Rubbers," *Jour. Chem. Phys.*, XIV (1946), 93; C. Mack, "Plastic Flow, Creep, and Stress Relaxation," *Jour. Appl. Phys.*, XVII (1946), 1086, 1093, 1101.

77. J. A. Ewing and W. Rosenhain, "Experiments in Micrometallurgy–Effects of Strain," *Phil. Trans. Roy. Soc., A*, CXCIII (1900), 353.

of the material between the slip bands failed to reveal any deformation whatsoever. Deformation through isolated slip bands rather than homogeneous deformation appears, in fact, to be characteristic of all crystalline materials. As will be discussed later in some detail, the evidence is very strong that the material in a previously formed slip band is, at least temporarily, viscous in the sense of obeying equation (209) rather than equation (210). In polycrystalline materials each slip band is confined to a single grain. It would therefore appear as if each slip band behaved as an isolated viscous region and thereby gave rise to anelastic effects. The properties of slip bands are hence very pertinent to a study of the anelasticity of metals.

It is, in fact, the viscous nature of slip bands which renders the yield stress of pure metals so low, two to three orders of magnitude below the theoretical yield stress of e_0E, where e_0 is of the magnitude of 0.1. As previously mentioned, an isolated viscous region will result in shear-stress concentration. Just in front of the spearhead of an advancing slip band the local shear stress is given by[78]

$$\text{Local shear stress} \simeq \sqrt{\frac{L}{r}}\ \text{over-all shear stress}\,,$$

where L is the length of the slip band and r the radius of curvature of its spearhead. According to this equation, the larger the slip band, the lower need be the over-all stress for continued propagation. The relation between critical stress and slip-band size is corroborated by two observations. First, the formation of a slip band appears to be a cataclysmic process.[79] Once the stress is sufficiently high to cause propagation, one would expect the slip band to grow at an increasing rate. Second, observations upon the proportional limits of copper and copper-base alloys[80] are in agreement[81] with the theoretical prediction based upon equation (211) that the stress at which yielding is first observed is inversely proportional to the grain size, i.e., to the maximum value of L.

The concept of slip bands propagating under their own stress concentration encounters the same difficulty as does the concept of cracks propagating under their own stress concentration. This concept is powerless to describe how the slip bands originate; for, until they attain a certain critical size, which is smaller, the higher the stress, their stress concentration

78. Inglis, *op. cit.*

79. I. W. Obreimow and L. W. Schubnikoff, *Zeitschr. f. Phys.*, XLI (1927), 907.

80. R. A. Wilkins and E. S. Bunn, *Copper and Copper Brass Alloys* (New York: McGraw-Hill Book Co., Inc., 1943).

81. C. Zener, "A Theoretical Criterion for the Initiation of Slip Bands," *Phys. Rev.*, LXIX (1946), 128.

is not sufficiently high to cause self-propagation. Some other mechanism of deformation must exist which leads eventually to the initiation of slip bands. The most striking evidence for pre-slip-band deformation is to be found in the experiments upon α-brass[82] and α-iron[83] single crystals. In these crystals the application of a load may be accompanied by no immediate observable deformation. Only after an "incubation" period of the order of minutes or hours does visible deformation start, and it then proceeds at an increasing rate. Less pronounced incubation periods have been found in commercially pure zinc single crystals.[84] It is possible that during the incubation period small slip bands are being formed through the movement of lattice defects, such as dislocations. The concept of dislocations was originated by Orowan[85] and Polanyi[86] and greatly extended by Taylor,[87] in order to explain why metals yield at a stress several orders of magnitude less than the theoretical yield stress. Although dislocations have been the subject of considerable theoretical investigation, a review of which has been given by Seitz and Read,[88] it is not clear just what is their role in plastic deformation, nor is it clear how they can initiate slip bands. Polycrystalline lead, freshly quenched from an anneal at 120°–130° C., shows an initially increasing creep rate,[89] a phenomenon which appears to be related to the incubation periods in α-brass and α-iron crystals.

From the above discussion we conclude that deformation proceeds primarily through the propagation of slip but that another mechanism of deformation also exists which, although contributing a negligible amount to the total deformation, is nevertheless essential to the initiation of slip bands. Before we leave this subject, it will be pertinent to review other experimental evidence which indicates deformation by some mechanism other than slip bands. In his experiments upon tin, Chalmers[90] has found

82. H. L. Burghoff and C. H. Mathewson, "Time and Temperature Effects in the Deformation of Brass Crystals," *Trans. A.I.M.E.*, CXLIII (1941), 45.

83. M. Gensamer and R. F. Mehl, "Yield Stress of Single Crystals of Iron under Static Load," *Trans. A.I.M.E.*, CXXXI (1938), 372.

84. T. A. Read and E. P. T. Tyndall, "Internal Friction and Plastic Extension of Zinc Single Crystals," *Jour. Appl. Phys.*, XVII (1946), 713.

85. E. Orowan, "Zur Kristallplastizität. III," *Zeitschr. f. Phys.*, LXXXIX (1934), 634.

86. M. Polanyi, "Über eine Arte Gitterstörung die einen Kristallplastische machen konnte," *Zeitschr. f. Phys.*, LXXXIX (1934), 660.

87. G. I. Taylor, *Proc. Roy. Soc., London,* CXLV (1934), 362.

88. F. Seitz and T. A. Read, "Theory of the Plastic Properties of Solids," *Jour. Appl. Phys.*, XII (1941), 100, 170, 470, 538.

89. J. N. Greenwood and H. K. Worner, "Types of Creep Curves Obtained with Lead and Its Dilute Alloys," *Jour. Inst. Metals*, LXIV (1939), 135.

90. B. Chalmers, *Proc. Roy. Soc., London*, CLVI (1936), 427; and *Jour. Inst. Metals*, LXI (1937), 103.

that at very low stresses the initial rate of creep is proportional to the applied stress but that the total extension produced by this creep is limited and is independent of the applied stress. It looks as though his specimens contained dislocations which moved at a rate proportional to the stress but that they could move only a definite distance, after which they were presumably stopped by some sort of barrier.

Plastic deformation usually increases anelastic effects. Examples have been reported, however, in which the prior application of a stress decreases the anelastic effects. In his experiments upon single and multiple crystals, Found[91] discovered that the internal friction at low-stress amplitudes could frequently be reduced by the prior application of a stress below the elastic limit and that the decrease was proportional to the applied stress. These experiments might be interpreted in a manner similar to that used in Chalmer's experiments. Found's specimens may have initially contained dislocations which contributed to the internal friction. A steady stress presumably moved these dislocations as far as a barrier, to which they appear to become attached and hence no longer contribute to the internal friction.

Independent evidence that deformation by slip bands is preceded by another type of deformation may be found in experiments upon rock salt. Here the yield stress—the stress necessary for the initiation of slip bands—is very dependent upon purity, the addition of only 0.001 mole per cent of $PbCl_2$ increasing the yield by a factor of 3.[92] On the other hand, the initiation of deformation in ionic crystals may also be detected[93] by the introduction of "color centers." The yield stress defined in terms of these color centers is, however, found to be quite insensitive to the degree of purity.[94] It looks as though the color centers introduced by stressing are the forerunners of slip bands and that the hardening influence of impurities is related to the inhibition which they exert in the transition of these color centers into slip bands. Although color centers have been extensively studied,[95] their relation to plastic deformation is at present unknown, aside from the fact that they are introduced by application of stresses.

91. G. H. Found, "Internal Friction of Single Crystals of Brass, Copper, and Aluminum," *Trans. A.I.M.E.*, CLXI (1945), 120.

92. F. Blank and A. Smekal, *Naturwiss.*, XVIII (1930), 306.

93. A. Smekal, *International Kongress Photographie* (Dresden, 1931), p. 34.

94. E. Poser, *Zeitschr. f. Phys.*, XCI (1934), 593.

95. N. F. Mott and R. W. Gurney, *Electronic Processes in Ionic Crystals* (Oxford: Oxford University Press, 1940), chap. iv.

b) RELATION OF PREVIOUSLY FORMED SLIP BANDS TO MECHANICAL PROPERTIES OF METALS

i. *Hysteresis.*—During the last century it was observed that plastic deformation introduces hysteresis loops in stress-strain curves.[96] Thus the stress-strain curve in annealed mild steel is straight and reversible, provided that the yield stress is not exceeded. Once the yield stress is exceeded, with accompanying plastic deformation, the stress-strain curve is no longer reversible, even at stress levels below the original yield stress, as illustrated in Figure 44. It was soon discovered that this hysteresis disappears with aging,[97] slowly at room temperature, within a few minutes at 100° C. Rosenhain[98] early recognized that this hysteresis and recovery could be interpreted in terms of the then recently discovered slip bands. He pointed out that the slip bands behave as if the material therein had an initially amorphous structure with corresponding viscous mechanical properties and as if this material gradually acquired the crystalline structure of the surrounding undeformed matrix.

Unfortunately, considerable confusion has arisen through the association of Rosenhain's concept of viscous-like slip bands with Beilby's[99] concept of amorphous layers. Beilby also considered the slip band material to be amorphous, but he assumed this amorphous structure to strengthen rather than to weaken the slip bands, and he did not believe that the material gradually acquired the crystalline structure of the surrounding matrix. This confusion has led to a gradual discrediting of Rosenhain's concept of viscous-like slip bands. Alternative theories have therefore been advanced to interpret the hysteresis introduced by plastic reformation. One current view is that the hysteresis is due to residual stresses introduced by plastic deformation. While such residual stresses could give rise to hysteresis at high stress levels, such hysteresis would vanish as the stress level was lowered. The hysteresis, or internal friction, approaches, however, a constant value as the stress level is lowered. Another argument contrary to the view point of residual stresses is that there is no evidence that the residual microscopic stresses are eliminated by low-temperature anneals at 100° C. Another current view is that the hysteresis introduced by plastic deforma-

96. J. Ewing, "On the Measurement of Small Strains in the Testing of Materials and Structures," *Proc. Roy. Soc., London*, LVIII (1895), 123.

97. J. Muir, "The Recovery of Iron from Overstrain," *Phil. Trans. Roy. Soc., London*, CXCIII (1900), 1; E. Coker, "On the Effect of Low Temperature on the Recovery of Overstrained Iron and Steel," *Phys. Rev.*, XV (1902), 107.

98. W. Rosenhain, "Deformation and Fracture in Iron and Steel," *Jour. Iron and Steel Inst.*, LXX (1906), 189.

99. G. Beilby, "The Hard and Soft States in Metals," *Phil. Mag.*, VIII (1904), 258.

tion arises through the partial mobility of dislocations introduced by deformation. As previously mentioned, it looks as though all deformation is confined to localized slip bands. The dislocations, if present, must then be within the slip bands themselves. However, a high concentration of dislocations would impart to the slip bands a viscous-like behavior. The viewpoint of a high concentration of dislocations within slip bands is therefore identical with Rosenhain's viewpoint of viscous-like behavior.

The development of electronic methods of measuring internal friction at low stress amplitudes has led to a considerable number of experiments[100] on the influence of plastic deformation. All these experiments confirm the

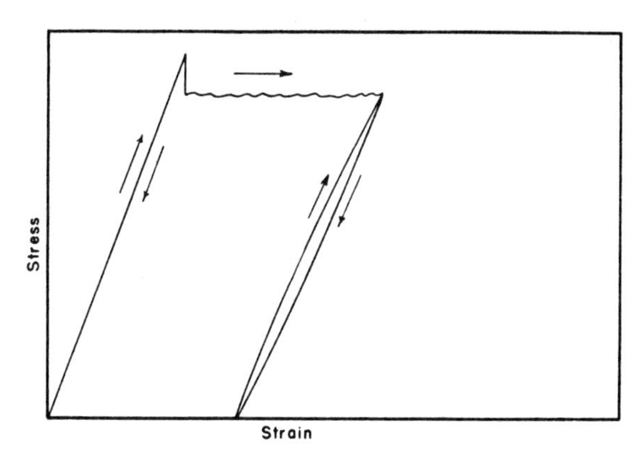

FIG. 44.—Example of hysteresis introduced by plastic deformation

early observations that plastic deformation introduces hysteresis and that this hysteresis is removed by low temperature anneals of from 100° to 200° C. In his work upon the effect of stress amplitude, Read[101] has found that, although the internal friction introduced by elastic deformation is independent of stress amplitude at very low stress levels, it increases rapidly with stress at comparatively low stress levels. Vibration at the higher stress levels does not affect the internal friction measured at the lower stress levels, i.e., internal friction is a single-valued function of stress

100. F. Förster and W. Köster, "Elastizitäts-Modul und Dämpfung in Abhänigkeit von Werkstoffzustand," *Zeitschr. f. Metallk.*, XXIX (1937), 116; W. Köster and K. Rosenthal, "Die Änderung von Elastizitäts-Modul und Dämpfung bei der Verformung und Rekristallisation von Messing," *Zeitschr. f. Metallk.*, XXX (1938), 343; W. Köster, "Elastizitätsmodul und Dämpfung von Aluminum und Aluminiumlegierung," *Zeitschr. f. Metallk.*, XXXII (1940), 282; T. A. Read, "The Internal Friction of Single Metal Crystals," *Phys. Rev.*, LVIII (1940), 371, and "Internal Friction of Single Crystals of Copper and of Zinc," *Trans. A.I.M.E.*, CXLIII (1941), 30; W. A. Lawson, "The Effect of Stresses on Internal Friction in Polycrystalline Copper," *Phys. Rev.*, LX (1941), 330; C. Zener, H. Clarke, C. S. Smith, "Effect of Cold Work and Annealing upon Internal Friction of Alpha Brass," *Trans. A.I.M.E.*, CXLVII (1942), 90.

101. *Op. cit.*

amplitude. Similar observations were also made by Norton[102] at much higher stress levels.

Considerable uncertainty exists at present regarding the relaxation spectrum associated with plastic deformation. On the one hand, many measurements[103] made at and in the vicinity of room temperature indicate that the internal friction introduced by plastic deformation is independent of the frequency of measurement and therefore that the relaxation spectrum is constant. On the other hand, one single piece of research[104] indicates that the internal friction associated with prior plastic deformation rises rapidly as the period of vibration is increased. By observations upon the elastic aftereffect of previously deformed iron at elevated temperatures, West[104] concluded that the internal friction increased with an increase in temperature and therefore, owing to the close relation of the effect of changes in temperature to that of changes in frequency of vibration, that it also increased with period of vibration. These conclusions are corroborated by current experiments by T. S. Kê.[105] It therefore appears that most information may be obtained regarding the mechanical properties of slip bands by experiments conducted at long relaxation times. Creep or stress relaxation experiments are particularly suited for obtaining this information.

ii. *Creep.*—As mentioned above, it appears that most information regarding the mechanical properties of the material within slip bands may be gathered from creep experiments at low stress levels and/or from the associated elastic aftereffect experiments. Numerous experiments have demonstrated that the elastic aftereffect is increased by cold working. These have been reviewed by Fromm.[106] The first attempt to gain an insight into the mechanical properties of the slip band material through such studies has been made by West,[104] who found that his observations upon plastically deformed iron at a series of temperatures could be correlated by ascribing to the viscosity of the slip-band material a heat of activation of 21,000 cal/mole. West's interpretation of his observations in terms of

102. J. Norton, "Changes in Damping Capacity during Annealing of Alpha Brass," *Trans. A.I.M.E.*, CXXXVII (1940), 49.

103. O. Föppl, "Drehschwingungsfestigkeit und Dämpfungsfähigkeit," *Ber. d. Werkstoffauschusses d. Ver. deutsch. Eisenhüttenleute*, No. 36, 1923; O. Föppl, "The Practical Importance of the Damping Capacity of Metals, Especially Steels," *Jour. Iron and Steel Inst.*, CXXXIV (1936), 393; A. Gemant and Willis Jackson, "The Measurement of Internal Friction in Some Solid Dielectric Materials," *Phil. Mag.*, XXIII (1937), 960; C. P. Contractor and F. C. Thompson, "The Damping Capacity of Steel and Its Measurement," *Jour. Iron and Steel Inst.*, CXLI (1940), 157; Zener, Clarke, Smith, *op. cit.*

104. W. A. West, *Trans. A.I.M.E.*, CLXVII (1946), 192.

105. T. S. Kê and C. Zener, "Structures of Cold Worked Metals as Deduced from Anelastic Measurements," *Symposium on Deformation of Crystalline Solids* (ONR, 1950), pp. 184–93.

106. *Op. cit.*

slip-band viscosity has been questioned by Kê,[107] who has suggested that the effects observed by West were probably associated with the high-temperature (150°–200° C.) relaxation in carbon- and nitrogen-bearing iron induced by plastic deformation,[108] the origin of which is not at present understood.

It is also commonly supposed that deformation necessarily raises the resistance to further deformation. This is the case only at lower temperatures. Extensive experiments, a review of which has been given by Burgers,[109] have shown that, at elevated temperatures, previously deformed specimens creep at a greater rate than do virgin specimens. Such accelerated creep presumably arises from the viscous behavior of the previously formed slip bands.

Once an insight has been gained, through creep experiments, into the viscous behavior of slip bands, it is anticipated, conversely, that this insight will give a better understanding of creep phenomena. In corroboration of this viewpoint, evidence will first be presented to show that the viscous behavior of slip bands is an important factor in the variation with temperature and with rate of deformation of the resistance to deformation. The dominating influence upon creep will then be discussed.

It is commonly supposed that the yield strength—the stress necessary to produce the first plastic deformation—decreases with a rise in temperature. When it is realized that the yield stress is commonly taken as the stress at which the permanent set is of the order of magnitude of 0.001 and that a considerable amount of strain hardening can occur during the first 0.001 strain, the commonly accepted viewpoint is open to considerable doubt. It might, on the other hand, be that the true yield stress, if such exists, is relatively insensitive to temperature and that the apparent rise of yield stress with a lowering of temperature is due merely to the increased rate of strain hardening at lower temperatures. Several experiments indicate that this viewpoint is indeed correct. In their careful observations upon the stress-strain curves of very pure single crystals of aluminum and silver, Miller and Milligan[110] have found that, as the temperature is raised from room temperature up to the recrystallization temperature, the stress actually increases, provided that the strains are less than 0.001. Only for larger strains does the stress decrease with an increase in temperature. At least for these crystals, the usual observation of decrease of yield stress

107. T. S. Kê, personal communication.

108. J. L. Snoek, *Physica*, VIII (1941), 711.

109. W. G. Burgers, *Handbuch der Metallphysik* (Leipzig, 1941), pp. 504–7.

110. R. F. Miller and W. E. Milligan, "Influence of Temperature on the Elastic Limit of Single Crystals of Aluminum, Silver, and Zinc," *Trans. A.I.M.E.*, CXXIV (1937), 229.

with increasing temperature is due merely to a decreased rate of strain hardening. In their work upon the initiation of yield in α-brass crystals, Burghoff and Mathewson[111] found that the critical resolved shear stress at which creep finally started, perhaps after a long incubation period, was, within their limits of error, independent of temperature from room temperature to 500° F. From the above experiments we are led to suspect that the commonly observed decrease in strength with increasing temperature is due primarily to viscous flow in previously formed slip bands, such flow thereby decreasing the measured strain hardening. The higher the temperature, the more rapid this viscous flow, and therefore the more is the strain hardening reduced. Again, the lower the rate of strain hardening the more dominant is this viscous flow, and therefore the more is strain hardening reduced. Only at the lowest temperatures is viscous flow within previously formed slip bands entirely suppressed and hence is strain hardening unaffected by such flow.

We shall now examine in detail the interrelation between strain hardening and creep, i.e., tensile deformation at constant load. A typical creep curve is reproduced in Figure 45. The rapidly decreasing creep rate in the first portion, in the primary creep range, is customarily attributed to strain hardening. We shall see below that equally important in the primary creep range is the variation of stress with strain rate, which, according to our present view, depends upon viscous flow in previously formed slip bands. The stationary creep rate of the secondary creep range is customarily believed to arise from an interaction of strain hardening and strain recovery. A criticism of this viewpoint is presented below, and an alternative viewpoint[112] is discussed. According to the latter viewpoint, the stationary secondary creep rate is dependent upon precisely the same metallurgical factors as those which govern the primary creep range. Finally, the rapidly increasing creep rate of the tertiary range is due either to the initiation of necking or to the formation of a multitude of small cracks within the specimen. The origin of these cracks is discussed under the section on stress relaxation across grain boundaries.

We shall now analyze the factors upon which primary creep depends. As discussed in detail by Hollomon,[113] the initial strain hardening of metals at constant strain rate is represented very well by the expression

$$S \sim e^m, \quad \dot{e} = \text{Constant}. \tag{213}$$

111. H. L. Burghoff and C. H. Mathewson, "Time and Temperature Effects in the Deformation of Brass Crystals," *Trans. A.I.M.E.*, CXLIII (1941), 45.

112. C. Zener and J. H. Hollomon, "Problems in Non-elastic Deformation of Metals," *Jour. Appl. Phys.*, XVII (1946), 69.

113. J. H. Hollomon, "Tensile Deformation," *Trans. A.I.M.E.*, CLXII (1945), 268.

The strain hardening exponent, m, is equal to the strain at which necking begins[114] and has been found to range from 0.06 for low-carbon steel to 0.7 for large grain size brass. On the other hand, at least in steel, stress varies with strain rate at constant strain, as[115]

$$S \sim \dot{e}^n, \quad e = \text{Constant} . \tag{214}$$

It is anticipated that the exponent n will be dependent upon temperature and somewhat upon the stress level. In steel at room temperature and below, n is of the order of magnitude of 0.01. Upon combining equations (213) and (214), we obtain

$$S \sim e^m \dot{e}^n . \tag{215}$$

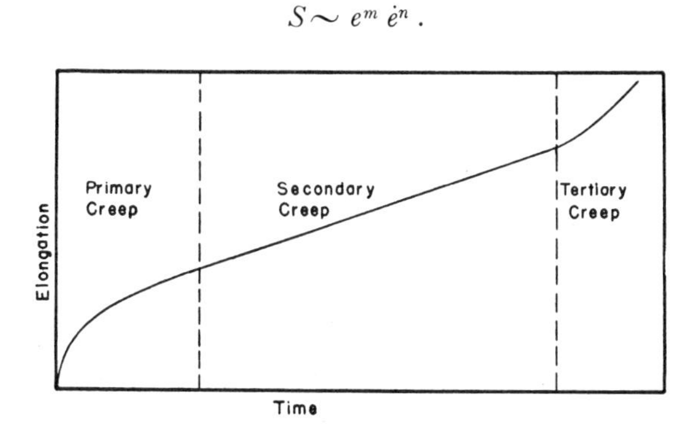

FIG. 45.—Typical creep curve (elongation versus time at constant load)

If we regard m and n as constants, the solution of equation (215) is

$$e = Ct^{n/(n+m)} , \tag{216}$$

where C is a constant. The creep rate, which is therefore given by

$$\dot{e} = bt^{-m/(n+m)} , \tag{217}$$

where b is another constant, approaches zero asymptotically. This conclusion is in agreement with the careful work of Andrade,[116] who showed that at constant stress the creep rate approached zero as $t^{-2/3}$. The rate at which the strain rate approaches zero is dependent both upon the strain hardening exponent m and upon the dependence of stress upon strain rate, a measure of which is given by the exponent n.

The customary creep experiments, to which Figure 45 refers, are not performed at constant stress but rather at constant load. As the specimen

114. *Ibid.*

115. C. Zener and J. H. Hollomon, "Effect of Strain Rate upon Plastic Flow of Steel," *Jour. Appl. Phys.*, XV (1944), 22.

116. E. N. Da C. Andrade, "On the Viscous Flow in Metals and Allied Phenomena," *Proc. Roy. Soc., London*, LXXXIV (1910), 1.

extends, the cross-section becomes smaller, and therefore the load increases. Although the increase of stress is negligible during the early part of the primary creep range, it is not necessarily negligible throughout the entire stress range. In order to find its influence, we must replace equation (215) by an equation in which load L appears rather than stress. If A refers to the cross-section area, then

$$L = AS\ .$$

If we neglect volume changes during creep,

$$A \sim \epsilon^{-e}\ ,$$

where ϵ is the base of the natural logarithm. The desired equation is therefore

$$L \sim \dot{e}^{n}\, e^{m} \epsilon^{-e}\ . \tag{218}$$

The creep rate is seen to be a minimum at the strain m and to increase thereafter. An example of the integral of equation (218) is given in Figure 46 for the particular case of $m/n = 10$. The creep rate is seen to have an essentially constant minimum value for a very long time. This result is a general consequence of equation (218) or of any other equation of the same general type; for, according to this or any similar equation, the creep rate is a minimum at some value of strain, say e_m. The creep rate will hence be essentially constant for strains some distance on either side of e_m. Since this essentially constant strain rate is also a minimum strain rate, a relatively long time will be spent in the region of essentially constant strain rate.

According to the above viewpoint, which was originally formulated by Ludwik,[117] the phenomenon of a secondary creep range is merely a consequence of the variation of stress upon strain and upon strain rate, and it is not necessary to invoke the additional concept of softening by annealing. Although it is known that strain hardening is removed at elevated temperatures, it is also known that the smaller the prior strain, the higher one must go in temperature to obtain strain recovery. No experiments have been performed to date to determine whether in the range of strain and temperature which are of practical importance to creep, appreciable strain recovery takes place.

iii. *Stress relaxation.*—As pointed out in chapter v, section B, precisely the same information may be obtained by stress relaxation experiments at low stress levels as by creep experiments at low stress levels. The viscous properties of slip band material may therefore be investigated by stress

117. P. Ludwik, *Elemente der technologischen Mechanik* (Berlin: Springer, 1909).

relaxation experiments. Conversely, any information so gained may be applied to reach an understanding of stress relaxation under practical conditions of operation.

Stress relaxation at constant strain is intrinsically a simpler physical phenomenon than is creep under ordinary conditions. The major fraction of an applied stress may be relieved through viscous flow in previously formed slip bands or along grain boundaries, and no generation of new slip bands is required. However, under ordinary creep conditions an increase in strain is presumably attended by the generation of new slip bands.

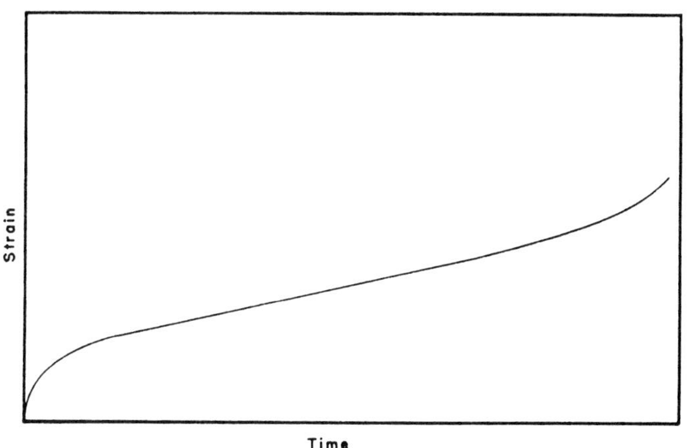

FIG. 46.—Theoretical creep curve according to equation (180) (drawn for case $m/n = 10$)

The results of stress relaxation experiments are usually plotted in a semilogarithmic fashion, i.e., stress versus logarithm of time. In their work upon lead, Trouton and Rankine,[118] presumably the first to study stress relaxation extensively, found that in such a plot the major part of their data lay upon a straight line. Thus in this region

$$S = a - b \log t \,, \tag{219}$$

where a and b are constants. Later investigators have found essentially the same result.

Equation (219) does not behave properly for small times, since $S(t)$ must approach a constant value $S(0)$ as t approaches zero. In order to remedy this defect, another constant is necessary. The new equation may be written as

$$S(t) = S(0)\, 1 - C ln \frac{t}{\tau + 1} \,. \tag{220}$$

118. F. T. Trouton and O. K. Rankine, "On the Stretching and Torsion of a Lead Wire beyond the Elastic Limit," *Phil. Mag.*, VIII (1904), 538.

This new equation, as well as the original equation (219), behaves improperly as t becomes very large. In order to avoid this difficulty, Trouton and Rankine have suggested still another modification, with, of course, the introduction of still another constant. This new equation is

$$S(t) = S(0)\,1 - Cln\left\{\frac{(t/\tau_1)+1}{(t/\tau_2)+1}\right\} \tag{221}$$

where

$$\tau_2 \gg \tau_1$$

and where

$$Cln\,\frac{\tau_2}{\tau_1} \leqslant 1\,. \tag{222}$$

The equality sign in equation (222) is to hold in case of complete stress relaxation at infinite time.

No satisfactory theoretical justification has been presented for an equation of the form of (219) for intermediate time ranges. A theoretical derivation has been given by Tobolsky and Eyring.[119] An assumption necessary to their argument is, however, that the creep rate is not linear in stress, but rather

$$\dot{e} \sim \sinh AS\,;$$

and their conclusion is valid only when AS is so large that the sinh function cannot be replaced by its argument. Equation (219) has been found valid, however, in the low stress range where creep rate is proportional to stress.

iv. *Moduli of elasticity.*—As early as 1907 Beilby[120] observed that plastic deformation decreased the moduli of elasticity and that this decrease was removed by a low-temperature anneal. Some confusion arose in later years from observations which indicated that, at least in certain cases, the moduli were increased. A clarification was finally given by Kawai[121] in 1930. Kawai found that a slight deformation always decreases the moduli and that in certain metals continued deformation leads finally to an increase in the moduli. A review of the literature on the subject has been given by

119. A. Tobolsky and H. Eyring, "Mechanical Properties of Polymeric Materials," *Jour. Chem. Phys.*, XI (1943), 125.

120. G. Beilby, *Proc. Roy. Soc., London*, LXXIX (1907), 463; *Jour. Inst. Metals*, VI (1911), 5.

121. T. Kawai, "The Effect of Cold Working on Young's Modulus of Elasticity," *Sci. & Tech. Repts., Tohoku Univ.*, XIX (1930), 209, and "On the Change of the Modulus of Rigidity in Different Metals Caused by Cold Working," *ibid.*, XX (1931), 681.

Elam.[122] Since this review, a thorough investigation by Köster[123] shows that the restoration of the original moduli on annealing closely parallels the removal of the internal friction introduced by plastic deformation.

A ready interpretation of the effect of plastic deformation in lowering the elastic moduli is to be found in the viscosity of the slip bands produced by this deformation. In fact, the decrease in elastic moduli is just one manifestation of this viscosity, of which hysteresis, creep, and stress relaxation are other manifestations. The simplest relation of the decrease of elastic modulus is with the stress relaxation function. Thus let $f(t)$ be the force necessary to cause a constant unit deformation to be suddenly applied at $t = 0$, and let $M(P)$ be the corresponding dynamic modulus measured at a period of vibration of P. Then from equation (67) we see that

$$M(P) = f(t)_{t \simeq P/8} .$$

This interpretation of the origin of the decrease in elastic moduli with plastic deformation leads to the prediction that no decrease in modulus would be observed if the measurements were carried out at very low temperatures. Such measurements have not been made.

The final rise in elastic moduli in certain metals due to continued plastic deformation presumably arises from reorientation effects which are associated with extensive deformation.

v. *Bauschinger effect.*—Annealed metals usually have the same yield stress in compression as in tension. Bauschinger[124] discovered in 1881 that this symmetry was destroyed by plastic deformation. If a metal is stretched in tension, its strength is raised with respect to a second application of a tensile stress but is lowered with respect to the application of a compressive stress. In fact, the metal yields plastically almost as soon as even a small compressive stress is applied.

The customary interpretation of the Bauschinger effect is in terms of the anisotropy of individual crystals with respect to their yield strength and their at least partially random orientation. When a tensile stress is slowly applied to a specimen, those grains yield first which are most favorably oriented for plastic deformation. As the stress is further increased, the stress increases less rapidly in the grains undergoing plastic deformation than in those grains which have remained elastic. Now, if the applied

122. C. F. Elam, *The Distortion of Metal Crystals* (London: Oxford University Press, 1936), p. 135.

123. W. Köster and K. Rosenthal, "Change in Elastic Modulus and Damping of Brass during Deformation and Recrystallisation," *Zeitschr. f. Metallk.*, XXX (1938), 345.

124. J. Bauschinger, *Zivilingenieur*, XXVII (1881), 299.

stress is removed, the grains which had undergone plastic deformation and hence had the smallest peak tensile stress will now be in compression, while the grains which had remained elastic will be in tension.

This customary interpretation of the Bauschinger effect can at most be a very rough approximation to the actual conditions in a plastically deformed specimen, since the deformation is not uniform throughout an individual grain. It does not, therefore, have much meaning to speak of the "residual stress" within an individual grain. It would appear more appropriate to speak in terms of individual slip bands. Just after a slip band has formed, the shear stress within it and in the elastic matrix on either side is to a great extent entirely relieved. If, now, the applied stress is removed, these regions will have shear stresses of the same magnitude but of a sign opposite to the original over-all stress. These regions will therefore yield under a very slight shear stress applied in the opposite direction. An interpretation of fatigue in metals may be based upon just these residual stresses.[125]

The observations of Sachs and Shoji[126] upon single crystals corroborate the above viewpoint. They found that single crystals show the Bauschinger effect to just as marked a degree as do polycrystalline specimens. It therefore appears that it is the residual stresses associated with individual slip bands rather than with individual grains which are responsible for the Bauschinger effect.

The foregoing interpretation leads to certain interesting predictions. The residual stresses across individual slip bands should be to a large extent removed by low-temperature annealing. The Bauschinger effect should likewise be so removed. On the other hand, it should be possible to reintroduce the Bauschinger effect without further plastic deformation. Application of a high stress at a slightly elevated temperature should reintroduce the residual stresses.

vi. *Fracture.*—Certain delayed fracture phenomena are most readily interpreted in terms of the gradual relaxation of shear stress across previously formed slip bands, with a resulting high accumulation of stress concentration at the edges of the slip bands. Most striking examples have been observed[127] by the author in armor-piercing projectiles. When an uncapped armor-piercing projectile strikes an armor plate under sufficiently severe conditions at normal incidence, the projectile penetrates in an in-

125. C. Zener and J. H. Hollomon, "Problems in Non-elastic Deformation of Metals," *Jour. Appl. Phys.*, XVII (1946), 69.

126. G. Sachs and H. Shoji, "Zug-Druchversuche an Messingkristallen," *Zeitschr. f. Phys.*, XLV (1927), 776.

127. C. Zener, "Anelasticity of Metals," *Trans. A.I.M.E.*, CLXVII (1946), 155.

tact condition but with the front portion having suffered a slight compression. After the projectile is recovered, its tip frequently flies off. An example is given in Figure 47. The delay between recovery and fracture varies from several minutes to several days. The force system acting upon the projectile is such that, after penetration, the central part of the front portion is under tension, the outer parts under compression. After recovery the residual tensile stress is presumably partially relaxed by shear stress relaxation across previously formed slip bands, this relaxation resulting in such high stress concentrations that small microcracks are eventually formed, which, in turn, are further opened by the residual tensile stress. The flying-off of the tip finally results from the acceleration by the com-

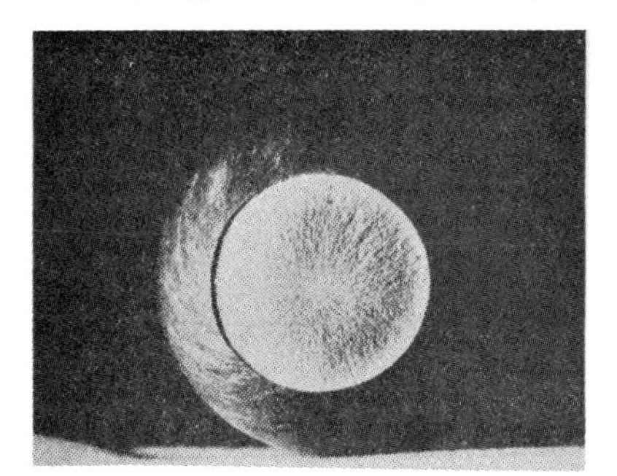

FIG. 47.—Example of tip of projectile which has been ejected after recovery of projectile

pressive stresses along the outer portion after the spreading of the crack throughout the central regions.

The rather frequent cases of delayed cracking in steel following quenching[128] may also be the result of the gradual building-up of high stress concentrations through stress relaxations across previously formed slip bands. A spectacular example has been described by Howe. "In the early days of making armor-piercing shells, spontaneous and violent aging rupture was so common that shells, after hardening, used to be stored for a considerable time in a room to which no one was admitted."[129]

III. STRESS RELAXATION ACROSS GRAIN BOUNDARIES

a) VISCOSITY OF GRAIN BOUNDARIES

Since about 1912 the existence of amorphous grain boundaries has been the subject of almost continual controversy. On the one hand, the evi-

128. S. Epstein, *Alloys of Iron and Carbon*, I (New York: McGraw-Hill Book Co., Inc., 1936), 323, 331; H. Carpenter and J. Robertson, *Metals* (London: Oxford University Press, 1939), pp. 711–12.

129. H. Howe and E. Groesbeck, "A Volute Aging Break," *Trans. A.I.M.E.*, LXII (1920), 522.

dence, as reviewed first by Rosenhain[130] and later by Jeffries and Archer,[131] is very convincing that grain boundaries behave as if they were amorphous. On the other hand, metallurgists have been reluctant to accept the concept of an amorphous phase. No controversy need arise, once it is realized that it is not necessary for any portion of the metal to be amorphous in order that the grain boundaries may behave in a viscous manner. It is necessary to assume only that the resistance to slipping of one grain over an adjacent grain obeys the laws commonly associated with amorphous materials rather than the laws associated with crystalline materials. Since the surface atoms of one grain cannot fit into the lattice positions of an adjacent grain, the binding across the interface of two grains may reasonably be expected to have the characteristics associated with amorphous materials.

The primary difference between the laws governing the mechanical behavior of amorphous and crystalline materials lies in the rapid increase in resistance to deformation of amorphous materials with respect to an increase in rate of deformation and to a decrease in temperature, compared with the relative insensitivity, with respect to these same variables, of the resistance to deformation of crystalline materials. Thus in amorphous materials of simple structure

$$\text{Shear stress} = \eta \times \text{shear strain rate} ,$$

where the coefficient of viscosity, η, varies with temperature as

$$\eta \sim e^{H/RT} ,$$

where R is the gas constant and H, the heat of activation, has a magnitude of several tens of thousands cal/mole. On the other hand, in the typical case of steel in the vicinity of room temperature, the shear stress varies as only about the one-hundredth power of the strain rate,[132] and the heat of activation, which describes the dependence upon temperature, is of the order of magnitude of only 100 cal/mole. It is therefore to be expected that, as the rate of deformation is lowered or as the temperature is raised, the resistance to slipping across the grain boundaries will be lowered with respect to the resistance to deformation within the interior of the grains. At sufficiently low rates of deformation and/or at sufficiently high temperatures, effects should be observed which are attributable only to a slip-

130. W. Rosenhain and D. Ewen, "Intercrystalline Cohesion in Metals," *Jour. Inst. Metals*, VIII (1912), 149; and W. Rosenhain, *Introduction to Physical Metallurgy* (London: Constable, 1915), pp. 255–64.

131. Z. Jeffries and R. Archer, *The Science of Metals* (New York: McGraw-Hill Book Co., Inc., 1924), chap. iv.

132. C. Zener and J. H. Hollomon, "Effect of Strain Rate upon the Plastic Flow of Steel," *Jour. Appl. Phys.*, XV (1944), 22.

ping across the grain boundaries. Such effects were first observed by Rosenhain and Humphrey.[133] When specimens of copper were polished and then extended at a slightly elevated temperature, those authors found that the grain boundaries became delineated by a relative displacement normal to the surface of adjacent grains. Such delineation does not occur at room temperature. Hanson and Wheeler[134] made a thorough study of grain boundary movement in aluminum. Here considerable relative movement is observed at the grain boundaries when the specimens are extended slowly at temperatures of 250° C. and above, but not when the specimens are extended rapidly an equivalent amount at room temperature. Andrade and Chalmers[135] found that a slight extension of cadmium at room temperature results in a decrease in electrical resistivity, while a like extension at liquid-air temperature raises the resistivity. It looks as though at room temperature the grains rotate in such a manner as to bring the axis of low resistivity closer to the tensile axis, without appreciable plastic deformation within the grains. Such a rotation could occur only by slipping along the grain boundaries. Moore, Betty, and Dollins[136] have obtained most striking evidence of viscous flow at grain boundaries. When specimens of lead are polished, scratched with parallel straight lines, and then pulled several per cent at room temperature, Moore *et al.* found that the scratches remained parallel in each grain but assumed different orientations in adjacent grains when the extension was slow and maintained the same orientation when the extension was rapid. Schumacher[137] has pointed out that discontinuities in the creep curves of lead are favored by low stresses and by large grains, and he has interpreted such discontinuities in terms of viscous grain-boundary slip. A review of the literature comparing the low creep resistance of fine-grained specimens in contrast to that of large-grained specimens has been given by Burgers.[138] The conclusive proof that grain boundaries do, in fact, behave in a truly viscous manner has recently been given by Kê, whose extensive work is referred to in the following sections.

133. W. Rosenhain and J. Humphrey, "The Crystalline Structure of Iron at High Temperatures," *Proc. Roy. Soc., London*, LXXXIII (1909), 200.

134. D. Hanson and M. A. Wheeler, "The Deformation of Metals under Prolonged Loading," *Jour. Inst. Metals*, XLV (1931), 229.

135. E. N. Da C. Andrade and B. Chalmers: "The Resistivity of Polycrystalline Wires in Relation to Plastic Deformation and the Mechanism of Plastic Flow," *Proc. Roy. Soc., London*, CXXXVIII (1932), 348.

136. H. Moore, B. Betty, and C. Dollins, *The Creep and Fracture of Lead and Lead Alloys* (University of Illinois Bull. 272 [1935]).

137. E. E. Schumacher, *Trans. A.I.M.E.*, CXLIII (1941), 176.

138. W. G. Burgers, *Handbuch der Metallphysik* (Leipzig, 1941), III, No. 2, 508–9.

b) RELATION OF THE VISCOSITY OF GRAIN BOUNDARIES TO THE MECHANICAL PROPERTIES OF METALS

i. *Hysteresis.*—Viscous flow is inevitably associated with the dissipation of mechanical energy into heat energy. Viscous flow at the grain boundaries should therefore contribute to the internal friction of metals subject to cyclic vibration. As long as the frequency of vibration is sufficiently high so that the shear stress across the grain boundaries relaxes only slightly during a half-cycle of vibration, the internal friction is expected to be higher, the higher the temperature and the greater the grain boundary area per unit volume, i.e., the smaller the grain size. These expectations have been confirmed by observations upon polycrystalline zinc[139] and brass[140] specimens at a variety of grain sizes and over a wide range of temperature. In all cases the internal friction varied inversely as the grain size and increased with rising temperature according to a heat of activation type of law.

If the individual grains are essentially equiaxed and the grain-size distribution is uniform, it is anticipated that only a fraction of the over-all stress can be relaxed by grain-boundary slip. The situation is analogous to the case of a jigsaw puzzle, in which the over-all configuration possesses rigidity in spite of the fact that no shearing stresses exist between adjacent pieces. In this case a continued rise in temperature will not lead to a continued rise in the internal friction. Thus the energy dissipated across a single boundary may be written as

$$\text{Energy dissipated} \sim (\text{relative displacement}) \times (\text{shear stress}) .$$

At low temperatures the first factor is negligible, since here the viscosity is so high that no appreciable relative displacement of adjacent grains occurs during a half-cycle of vibration. At high temperatures the second factor is negligible, since here the viscosity is so low that the shear stress across all grain boundaries is essentially completely relaxed at all times. Only in an intermediate temperature range, where there is some relative displacement and some shear stress across the grain boundaries, is the dissipation of energy appreciable. An example of such an optimum temperature range has been demonstrated by Kê.[141] His results are shown in Figure 48.

The magnitude of the internal friction in the optimum temperature

139. A. Barnes and C. Zener, "Internal Friction at High Temperatures," *Phys. Rev.*, LVIII (1940), 87.

140. C. Zener, D. Van Winkle, and H. Nielsen, "High Temperature Internal Friction of Alpha Brass," *Trans. A.I.M.E.*, CXLVII (1942), 98.

141. T. S. Kê, "Experimental Evidence of the Viscous Behavior of Grain Boundaries in Metals," *Phys. Rev.*, LXXI (1947), 533.

range should be independent of grain size. This conclusion may be arrived at from two viewpoints. In the optimum temperature range the internal friction is proportional to the grain-boundary area per unit volume times the average relative displacement across a grain boundary. Now the grain-boundary area is inversely proportional and the relative displacement is directly proportional to the grain size. Their product is therefore independent of grain size. On the other hand, we may inquire into the total over-all stress relaxed by the grain-boundary flow, the over-all strain being maintained constant. The fraction of stress relaxed is a dimension-

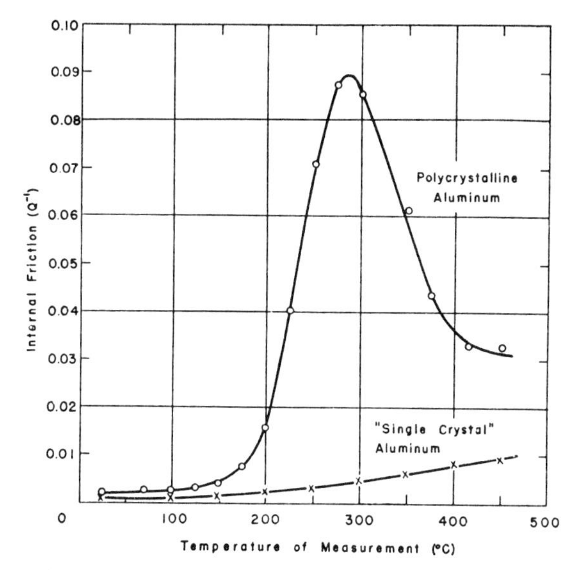

FIG. 48.—Variation of internal friction with temperature in polycrystalline and "single-crystal" aluminum (frequency of vibration = 0.8 cycles per second at room temperature). (After Kê.)

less quantity and cannot therefore depend upon grain size, since there is no other quantity having dimensions of length upon which the stress relaxation can depend, as long as the grain size is small compared with the smallest dimensions of the specimen. This nondependence of maximum internal friction upon grain size is illustrated in Figure 49 from further work of Kê.[142] It is seen that, as long as the grain size is small compared with the diameter of the specimen, the effect of an increase of grain size is merely to raise the temperature for the maximum internal friction.

An increase in the frequency of vibration has essentially the same effect as an increase in grain size, namely, a higher temperature is required for

142. T. S. Kê. "Stress Relaxations across Grain Boundaries in Metals," *ibid.*, LXXII (1947), 41.

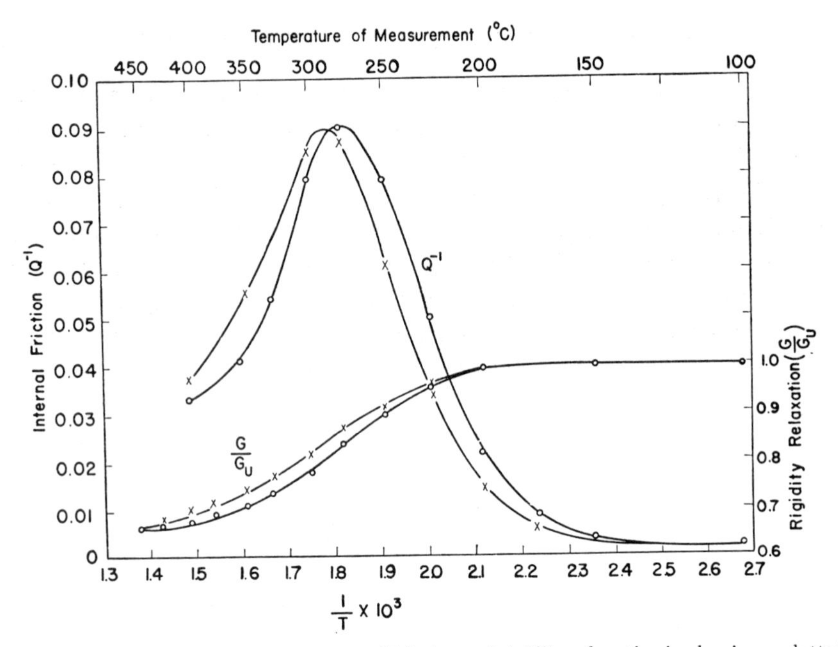

FIG. 49.—Effect of grain size on internal friction and rigidity relaxation in aluminum plotted against $1/T$ (frequency of vibration = 0.69 cycles per second at room temperature). Average grain diameter: ⊙, 0.02 cm.; ×, 0.04 cm. (After Kê.)

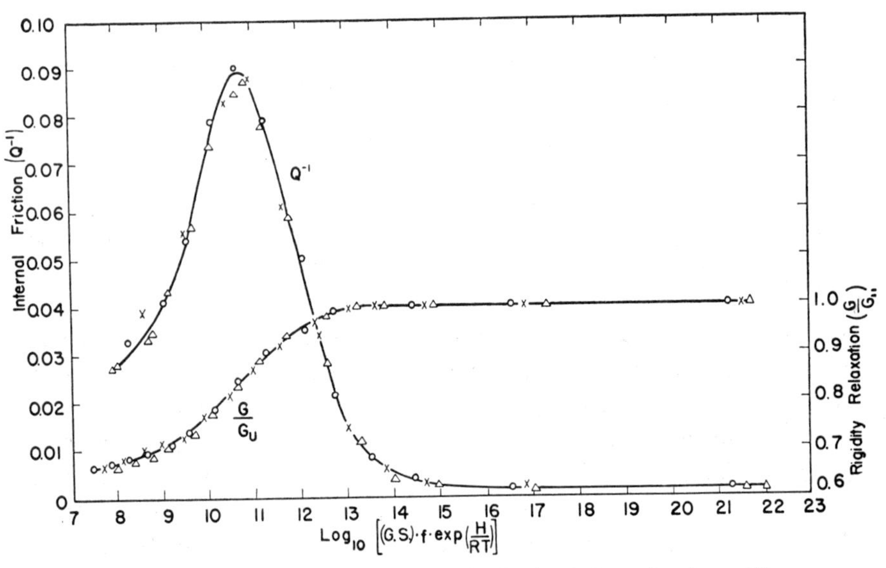

FIG. 50.—Internal friction and rigidity relaxation in aluminum as functions of the parameter (G.S.) $\times f \times \exp(H/RT)$, H = 32,000 calories per mole.

⊙ (G.S.) = 0.02 cm., f_{RT} (frequency of vibration at room temperature) = 0.69 cycles per second

× (G.S.) = 0.04 cm., f_{RT} = 0.69 cycles per second

△ (G.S.) = 0.02 cm., f_{RT} = 2.16 cycles per second

(After Kê.)

the same degree of relaxation. The interrelation of frequency, grain size, and temperature is beautifully demonstrated in Figure 50, taken from Kê.

Kê[143] has shown that the heat of activation associated with grain boundary slip is essentially identical to the heat of activation for self-diffusion and for creep, as is demonstrated in Table 13 taken from his work. Because of this correlation he has suggested that grain boundary slip involves the same mechanism as does volume diffusion.

ii. *Creep and creep recovery.*—Shear stress relaxation across grain boundaries under constant stress conditions must necessarily be accompanied by an extension, i.e., by creep. The creep rate will be smaller, the larger the grain size. This viewpoint is in accord with the general recognition[144] that increase in grain size favors an increased resistance to creep. As one example, in his work upon high-purity lead, McKeown[145] found that the minimum creep rate for a given load was essentially inversely proportional to the grain size and hence to the total grain area per unit volume. As another example, Burghoff[146] and collaborators found that the creep resistance of brass wire increased with an increase in grain size. If creep tests are not made sufficiently slowly, the usual correlation between grain size and creep rate will not be observed, as was, for example, the case in the experiments of Parker and Riisness.[147]

TABLE 13

DIFFERENT TYPES OF ACTIVATION ENERGY FOR DIFFERENT METALS

(In Calories/Mole)

Metal	Volume Diffusion	Grain Boundary Slip	Creep
α-brass	41,700	41,000	42,000
α-iron	78,000	85,000	90,000
Aluminum	37,500	34,500	37,000

143. T. S. Kê, "The Structure of Grain Boundaries in Metals," *Phys. Rev.*, LXXIII (1948), 267.

144. H. Carpenter and J. M. Robertson, *Metals* (London: Oxford University Press, 1939), p. 1415.

145. J. McKeown, "Creep of Lead and Lead Alloys. I. Creep of Virgin Lead," *Jour. Inst. Metals*, LX (1937), 201.

146. H. L. Burghoff, A. I. Blank, and S. E. Maddigan, "The Creep Characteristics of Some Copper Alloys at Elevated Temperatures," *Trans. A.S.T.M.*, XLII (1942), 668.

147. E. R. Parker and C. F. Riisness, "Effect of Grain Size and Bar Diameter on Creep Rate of Copper at 200° C," *Trans. A.I.M.E.*, CLVI (1944), 117.

An increase in creep rate with a decrease in grain size will be obtained only if the temperature is sufficiently high, as well as the rate of creep sufficiently low, that the grain boundaries manifest a smaller resistance to deformation than does the grain interior. Thus, if creep tests are made at a set of temperatures, a critical temperature will be found below which small-grained specimens are more creep resistant and above which large-grained specimens are more creep resistant. Examples have been cited by Clark and White.[148]

The relaxation of shear stress across grain boundaries must be accompanied by a readjustment of stresses within the interior of the grains. Such stress readjustment will result in residual stresses of such a sign as to produce slip across the grain boundaries in an opposite direction to that caused by the original load. In other words, these residual stresses will give rise to an at least partial creep recovery. Shear stress relaxation across grain boundaries as one possible source of creep recovery was recognized by Jeffries and Archer[149] in their discussion of the creep and marked creep recovery of a very fine-grained ternary eutectic. A demonstration that the creep occasioned by viscous slip across grain boundaries is essentially completely recoverable has been given by Kê,[150] whose observations are reproduced in Figure 51.

iii. *Stress relaxation.*—We have seen (p. 50) that, as long as the stress level is sufficiently low that the superposition principle is valid, the stress relaxation function is essentially the reciprocal of the creep function. One therefore anticipates stress relaxation at constant deformation under the same conditions as for creep and expects that viscous slipping at grain boundaries will be responsible for stress relaxation when, under the same conditions, it is responsible for creep. Numerous stress relaxation experiments have been conducted in recent years.[151] The tests have been made on commercial materials, with no attempt to understand the physical origin of the stress relaxation.

148. C. L. Clark and A. E. White, *The Working of Metals* (A.S.M., 1937), chap. on "Creep Characteristics of Metals."

149. Z. Jeffries and R. Archer, *The Science of Metals* (New York: McGraw-Hill Book Co., Inc., 1924), chap. iv.

150. T. S. Kê, "Experimental Evidence of the Viscous Behavior of Grain Boundaries in Metals," *Phys. Rev.*, LXXI (1947), 533.

151. N. L. Mochel: "Relaxation Tests on K20 Steel at 850° F," *Trans. A.S.M.E.*, LIX (1937), 453; J. Boyd, "Relaxation of Stress at Normal and at Elevated Temperatures," *Proc. A.S.T.M.*, XXXVII (1937), 218; C. C. Davenport, "Correlation of Creep and Relaxation Properties of Copper," *Trans. A.S.M.E.*, LX (1938), A-55; W. E. Trumpler, "Relaxation of Metals at High Temperatures," *Jour. Appl. Phys.*, XII (1941), 248; E. A. Davis, "Creep and Relaxation of Oxygen-free Copper," *Trans. A.S.M.E.*, LXV (1943), A-101; E. A. Rominski and H. F. Taylor, "Stress Relief and the Steel Casting," *Amer. Foundrymen's Assoc.*, LI (1944), 709.

The anticipated reciprocity of stress relaxation and of creep associated with grain-boundary slip has, in fact, been found by Kê.[150] Previously, West[152] had conducted stress relaxation experiments in iron and had interpreted his results in terms of grain boundary viscosity. These interpretations have been questioned[153] by Kê, who has observed that in iron a carbon content such as that contained in West's specimens is sufficient to block grain-boundary slip.

iv. *Elastic moduli.*—It has long been recognized that the elastic moduli decrease with an increase of temperature. In 1891 Sutherland[154] found that

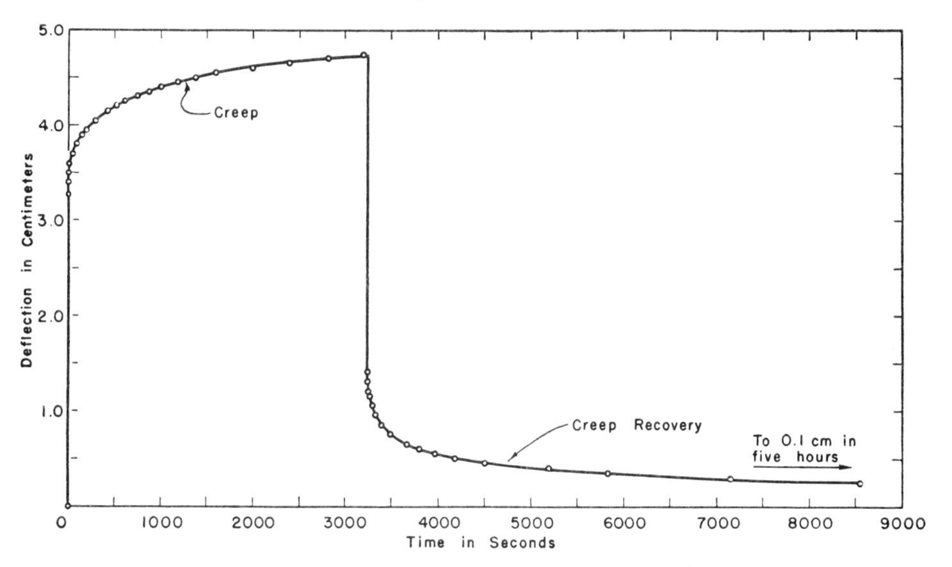

FIG. 51.—Creep under constant stress and creep recovery at 175° C. in polycrystalline aluminum. (After Kê.)

the data then existing for the rigidity modulus, G, of nonferrous metals were in agreement with the formula

$$\frac{G}{G_0} = 1 - \left(\frac{T}{T_m}\right)^2,$$

where G_0 is the rigidity modulus at the absolute zero and T_m is the melting temperature. Although his measurements extended to a T/T_m of only 0.75, Sutherland was sufficiently confident of the extrapolation to T_m to believe that the rigidity approaches zero as the melting temperature is

152. W. A. West, "Elastic After-effects in Iron Wires from 20° to 550° C.," *Trans. A.I.M.E.*, CLXVII (1946), 192.

153. T. S. Kê, personal communication.

154. W. Sutherland, "A Kinetic Theory of Solids, with an Experimental Introduction," *Phil. Mag.*, XXXII (1891), 31.

approached. Later observations by Jasper[155] were found to be in agreement with Sutherland's formula. The data for iron and steel, which have been reviewed by Sisco,[156] show the same general trend as Sutherland's formula. The concept that the rigidity modulus becomes essentially zero as the melting temperature is approached has received rather widespread acceptance, as may be seen from the discussion of Brillouin.[157]

Whether the observed rapid decrease of the rigidity modulus at high temperatures is a characteristic of the interior of the grains or is merely a manifestation of stress relaxation across grain boundaries can be decided only by a comparison of the temperature variation of the rigidity modulus of single crystals and of polycrystalline specimens. The work of Hunter and Siegel[158] upon a single crystal of NaCl, and of Siegel and Cummerow[159] upon a lead crystal, indicates that the latter is the case. Thus in the former example the rigidity modulus, C_{44}, decreased only 28 per cent as the specimen was raised from room temperature to just below melting temperature. The recent experiments of Kê[160] upon single crystalline and polycrystalline aluminum, shown in Figure 52, demonstrate conclusively that the usually observed rapid decrease of the rigidity modulus at high temperatures is, in fact, due to viscous slipping along the grain boundaries. The dynamic rigidity modulus of the single crystal decreased slowly in a linear manner as the temperature was increased to 400° C., while that of the polycrystalline specimen departed rapidly from this linear relation at about 200° C. Prior measurements by Birch and Bancroft[161] on the dynamic shear modulus of polycrystalline aluminum agree precisely with Kê's measurements on a single crystal up to 400° C., departing from a linear relation only above this temperature. Such a correlation is understood when cognizance is taken of the high frequency used by Birch and Bancroft, 8,000–5,000 cycles/sec in contrast to a frequency of about 1 cycle/sec used by Kê. The temperature at which grain boundary relaxation becomes appreciable is raised from about 250° to 450° C. by this increase in

155. T. M. Jasper, *Phil. Mag.*, XLVI (1923), 609.

156. F. T. Sisco, *The Alloys of Iron and Carbon*, Vol. II. *Properties* (New York: McGraw-Hill Book Co., Inc., 1937), pp. 438–42.

157. L. Brillouin, "On Thermal Dependence of Elasticity in Solids," *Phys. Rev.*, LIV (1938), 916.

158. L. Hunter and S. Siegel, "The Variation with Temperature of the Principal Elastic Moduli of NaCl near the Melting Point," *Phys. Rev.*, LXI (1942), 84.

159. S. Siegel and R. Cummerow, "On the Elasticity of Crystals," *Jour. Chem. Phys.*, VIII (1940), 847.

160. T. S. Kê, *Phys. Rev.*, LXX (1946), 105A; LXXI (1947), 533.

161. F. Birch and D. Bancroft, "Elasticity of Crystals," *Jour. Chem. Phys.*, VIII (1940), 641.

frequency. Similar observations have been reported by Kê for grain boundary relaxations in the case of α-brass[161a] and in the case of α-iron.[161b]

A computation has been made[162] of the maximum effect that stress relaxation across grain boundaries can have upon Young's modulus in the particular case in which the grains are regular polyhedra of identical size.

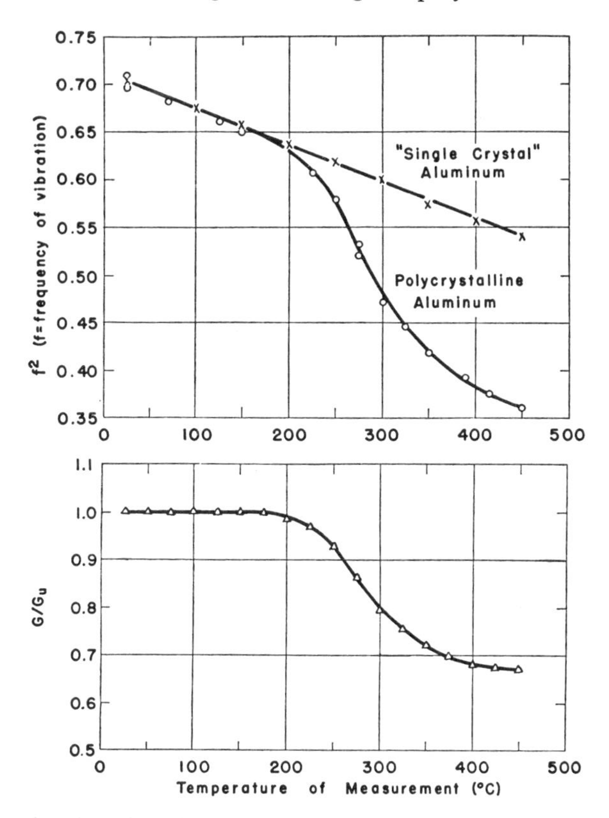

Fig. 52.—Relaxation of rigidity modulus in polycrystalline aluminum. (After Kê)

The ratio of the completely relaxed modulus, E_R, to the unrelaxed modulus, E_U, was found to vary from 0.62 to 0.68 over the usual range of Poisson's ratio of 0.3–0.4. From the analysis on pages 56–57 we saw that the relaxation in the shear modulus is about 15 per cent greater than the relaxation in the Young's modulus, so that the ratio G_R/G_U should lie within the range 0.56–0.63 for the usual range of values of Poisson's ratio. If the

161a. T. S. Kê, "Viscous Slip along Grain Boundaries and Diffusion of Zinc in Alpha Brass," *J. Appl. Phys.*, XIX (1948), 285.

161b. T. S. Kê, "Anelastic Properties of Iron," *Trans. A.I.M.E.*, (in press) (1948).

162. C. Zener, "Theory of Elasticity of Polycrystals with Viscous Grain Boundaries," *Phys. Rev.*, LX (1941), 906.

grains are not regular polyhedra or of uniform grain size, the relaxation will be greater, and G_R/G_U may approach zero. The experimental value of the ratio G_R/G_U in equiaxed aluminum may be seen from Figure 52 to be in complete agreement with the theoretical value.

v. *Fracture.*—As discussed on pages 126–30, the relaxation of shear stress in an isolated region will lead to high stress concentration, which may result in the formation of small microcracks. An example of how shear stress relaxation across a grain boundary may lead to a microcrack is illustrated in Figure 53. Here tensile stress is taken as applied along an axis normal to the grain boundary, BB^1. This boundary is intersected by another grain boundary, AA^1, across which the tensile stress has a shear

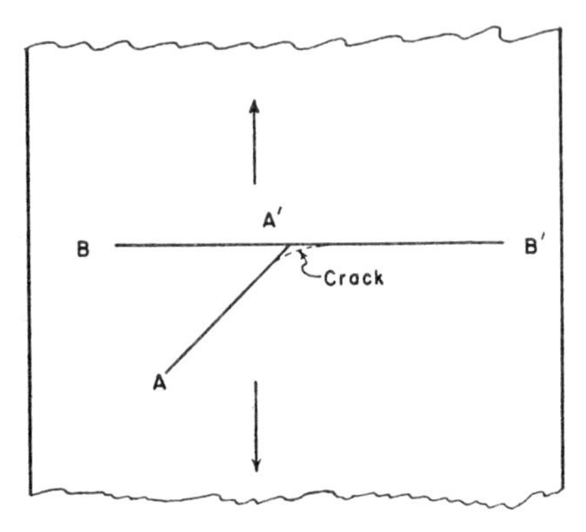

Fig. 53.—Illustration of how microcracks can be formed by shear stress relaxation along the grain boundaries.

stress component. Relaxation of shear stress across the boundary AA^1 will result in an extremely high tensile stress in the corner AA^1B^1. When this tensile stress reaches the intrinsic fracture strength, a small microcrack will form. Further slip along the plane AA^1 will result in a spread of the microcrack until the crack has such a size that it may propagate under its own stress concentration. Further propagation need not then wait for further slip along the plane AA^1 but proceeds at a great rate. Fracture of the specimen ensues.

The role played by the viscosity of grain boundaries at elevated temperatures in the initiation of creep was first recognized by Rosenhain[163] and his colleagues in their work upon the deformation of γ-iron at high temperatures. When pulled under the usual test conditions, this iron mani-

163. W. Rosenhain and J. C. W. Humphrey, "The Tenacity, Deformation, and Fracture of Soft Steel at High Temperatures," *Jour. Iron and Steel Inst.*, LXXXVII (1913), 219; W. Rosenhain and D. Ewen, "Intercrystalline Cohesion in Metals," *Jour. Inst. Metals*, No. 2, 1912, p. 149; No. 2, 1913, p. 119.

fested fair ductility and fractured in a fibrous manner. When extended very slowly at the same temperature, thereby allowing time for viscous slip, the iron fractured in an intercrystalline manner, with hardly any prior plastic deformation. The occurrence of this type of intercrystalline failure in a wide range of metals was later discussed by Rosenhain and Archbutt,[164] who demonstrated that such failure is facilitated by small plane-grain boundaries, grain boundaries that can slip without mechanical hindrance. Season cracking in brass was ascribed by them to grain boundary slip. More recently Hanson and Wheeler[165] have made a study of the role that grain boundaries play in the fracture of aluminum under creep conditions. They found that at elevated temperatures cracks invariably started at the grain boundaries under slow extension and propagated into the grains only just before fracture, when the rate of extension was large and hence when the resistance to deformation of the viscous grain boundaries became greater than that of the grain interior. As spectacular evidence of crack formation at grain boundaries, they showed that under creep conditions the density of polycrystalline specimens decreased appreciably, while that of single crystals remained constant. A beautiful example of intercrystalline cracks in copper found by slow extension at elevated temperatures has been given by McAdam[166] and is reproduced as Figure 54.

It is well recognized that the conditions which favor viscous flow at grain boundaries—high temperature and low rates of extension—also favor low ductility and that the resulting essentially brittle failure is even more deleterious than creep under service conditions. As pointed out by Carpenter and Robertson,[167] those factors which increase the resistance of metals to creep do not necessarily increase their resistance to brittle failure by the same amount.

IV. STRESS RELAXATION ACROSS TWIN INTERFACES

Twins induced by stress usually form by a discontinuous process, their appearance being accompanied by a characteristic clicking sound. Once a twin band has formed, however, its boundaries may move in a continuous manner. A well-known example is furnished by recrystallized α-brass. In

164. W. Rosenhain and S. L. Archbutt, "On the Intercrystalline Fracture of Metals under Prolonged Application of Stress," *Proc. Roy. Soc., London,* XCV (1919–20), 55.

165. D. Hanson and M. A. Wheeler, "The Deformation of Metals under Prolonged Loading. I. The Fracture and Flow of Aluminum," *Jour. Inst. Metals,* XLV (1931), 229.

166. D. J. McAdam, G. W. Geil, and D. H. Woodard, "Influence of Temperatures and of Strain Rate on the Mechanical Properties of Monel Metal and Copper," *Proc. A.S.T.M.,* XLVI (1946), 902.

167. *Op. cit.,* pp. 456, 462.

this metal each crystal is traversed by several twin bands, the number of twin bands, and the relation of their width to the diameter of the individual grains being essentially independent of grain size. Apparently, as the grains grow, the width of the twin bands likewise grows by a continuous displacement of the twin interface boundaries.

As may be seen by reference to Figure 42, a shear stress across a twin interface tends to induce a movement of this interface. Thus in the case illustrated, a mass movement of the right-hand twin member in the direction of the applied traction is accompanied by the movement of the twin

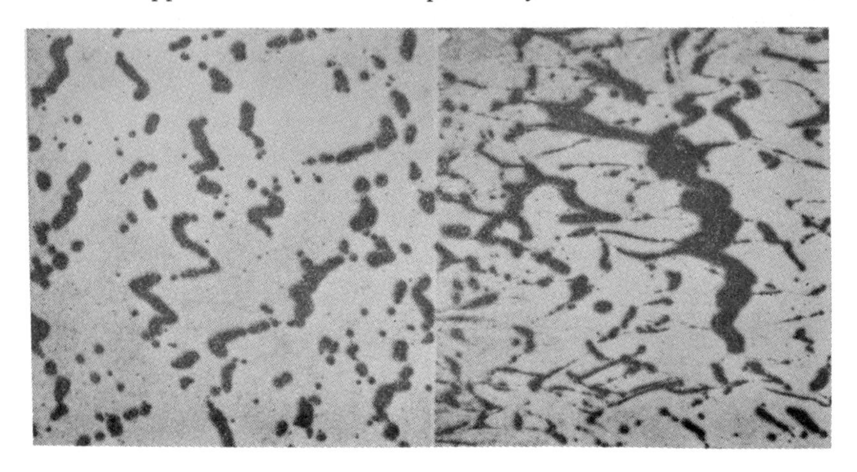

Unetched (×250) Etched with a mixture of equal parts of concentrated nitric and glacial acetic acids (×250)

Fig. 54.—Longitudinal sections after creep tests of cold-drawn monel metal

interface from AA' to BB'. Conversely, twin interfaces will tend to adjust their positions in such a manner as to minimize the shear stress acting across them. This continual striving of the twin interfaces to minimize the shear stresses will give rise to characteristic anelastic effects.

The anelasticity of copper manganese alloys (~90 per cent Mn) has been studied by F. Worrell,[168] and is thought to arise from the stress-induced movement of the twin interfaces of this alloy. The temperature variation of the internal friction of a typical specimen is given in Figure 55. In the low-temperature range the motion of the twin interfaces is so slow that no appreciable displacement occurs during a half-cycle of vibration, and so the internal friction is very low. On the other hand, in the high-temperature range the twin interfaces are so mobile that the shear

168. F. Worrel, "Twinning in Tetragonal Alloys of Copper and Manganese," *Jour. Appl. Phys.*, XIX (1948), 929; A. V. Siefert and F. Worrel, "The Role of Tetragonal Twins in the Internal Friction of Copper Manganese Alloys," *Jour. Appl. Phys.*, XXII (1951), 1257.

stress is completely relieved at all times, so the internal friction is again very low. Only in the intermediate-temperature range, where both the displacement and the net shear stress are appreciable, is the internal friction large.

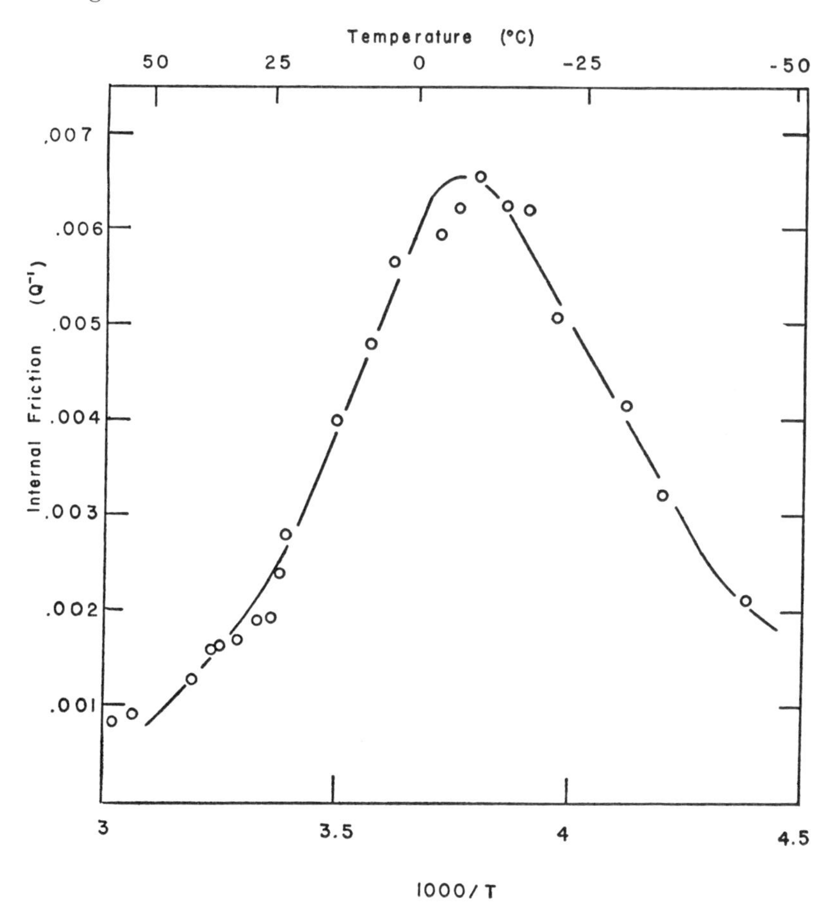

FIG. 55.—Typical variation of internal friction of CuMn alloy with temperature. This internal friction is supposedly due to movement of twin interfaces. (After Worrell.)

The alloys of manganese with f.c.c. metals are especially suited to a study of the laws governing the movement of twin interfaces. It is commonly thought that the high-temperature γ-phase of manganese is face-centered tetragonal (f.c.t.) and that the tetragonality ratio, c/a, increases gradually from 0.934 for pure Mn to unity as the concentration of a f.c.c. metal is increased. As yet no high-temperature X-ray structures have been determined, the f.c.t. structure being deduced from measurements on quenched specimens. The highly twinned nature of the CuMn alloys, first discovered by Worrell and illustrated in Figure 56, suggest that the high-temperature structure is, in fact, cubic and that the tetragonality is induced by a diffusionless transition during quenching.[169] According

FIG. 56.—Example of finely spaced (101) twins of face-centered tetragonal CuMn alloy superposed upon twins characteristic of a face-centered cubic lattice. (After Worrell.)

ing to this viewpoint, heavy twinning must arise in order that high residual stresses may be avoided. Whatever the origin of the twins in the manganese alloys, they are of unusual theoretical interest because of the smallness of the atomic movements associated with the motion of the twin interfaces. Thus, in the interface movement illustrated in Figure 40, the shear strain between the planes AA' and BB' varies from 0.066 in pure γ-manganese to less than 0.015 in γ-manganese containing about 18 per

169. This suggestion of the cubic nature of the high-temperature phase has been verified by U. Zwicker, *Zeitschr. Metallk.*, XLII (1951), 246.

cent copper. In the latter case the atomic displacements associated with the movement of a twin interface by one interatomic distance are of the same order of magnitude as the amplitude of vibration at the absolute zero temperature. The same type of twinning as found by Worrel in CuMn has also been found in CuAn,[170] FePt,[171] barium titanate,[172] CoPt[173], CrMn,[174] and in InTh.[175]

A second example of stress-induced movement of twin interfaces occurs in the case of magnetic domains. In a stress free specimen of iron, the local magnetization is along one of the principal axes, resulting in local tetragonality through magnetostriction. The boundary of two adjacent domains is along a (110) plane[176] and thus may be regarded as a twin interface, the two domains joining with no elastic distortion whatsoever. The application of a stress visibly moves these interfaces, and the time lag in this movement gives rise to anelastic effects. Since the tetragonality in the magnetic domains is very small, the c/a ratio being 1.000032,[177] a very small amount of stress relaxation results in a large movement of the twin interface.

170. J. L. Haughton and R. J. M. Payne, *Jour. Inst. Metals*, XLVI (1931), 457.

171. H. Lipson, D. Shoenberg, and G. V. Stuport, *Jour. Inst. Metals*, LXVII (1941), 333.

172. B. Matthias and A. von Hippel, *Phys. Rev.*, LXXIII (1948), 1378.

173. E. Gebhart and W. Köster, *Zeitschr. f. Metallk.*, XXXII (1940), 253; J. B. Newkirk, A. H. Geisler, D. L. Martin, and R. Smoluchowski, *Trans. A.I.M.E.*, CLXXXVIII (1950), 1249.

174. S. J. Carlile, J. W. Christian, and W. Hume-Rothery, *Jour. Inst. Metals*, LXXVI (1949), 169.

175. L. Guttman, *Trans. A.I.M.E.*, CLXXXVIII (1950), 1472.

176. J. H. Williams, *Phys. Rev.*, LXXI (1947), 646.

177. F. Bitter, *Introduction to Ferromagnetism* (New York: McGraw-Hill Book Co., Inc., 1937), pp. 250–54.

INDEX

INDEX